Do Lemmings Commit Suicide? · · · · · ·

Frontispiece. A collared (or varying) lemming taking a non-suicidal leap onto a blanket six inches below. Photo by Ian Lindsay.

Do Lemmings Commit Suicide?

· · · · · ·

Beautiful Hypotheses and Ugly Facts

Dennis Chitty

New York   Oxford
OXFORD UNIVERSITY PRESS
1996

Oxford University Press

Oxford   New York
Athens   Auckland   Bangkok   Bombay
Calcutta   Cape Town   Dar es Salaam   Delhi
Florence   Hong Kong   Istanbul   Karachi
Kuala Lumpur   Madras   Madrid   Melbourne
Mexico City   Nairobi   Paris   Singapore
Taipei   Tokyo   Toronto

and associated companies in
Berlin   Ibadan

Copyright © 1996 by Oxford University Press, Inc.

Published by Oxford University Press, Inc.,
198 Madison Avenue, New York, New York 10016

Oxford is a registered trademark of Oxford University Press

Library of Congress Cataloging-in-Publication Data
Chitty, Dennis.
    Do lemmings commit suicide? : beautiful hypotheses and ugly facts
/ Dennis Chitty.
       p.   cm.
    Includes bibliographical references (p.   ) and index.
    ISBN 0-19-509785-8; ISBN 0-19-509786-6 (paper)
    1. Lemmings.   2. Lemmings—Behavior.   3. Rodent populations.
I. Title.
QL737.R666C48   1996   95-22673
599.32'33—dc20

9 8 7 6 5 4 3 2 1

Printed in the United States of America
on acid-free paper

To Helen 1910–1987

Foreword

Some stories are so good that we can't stop telling them even after we know that they have little or no foundation in fact. Female praying mantises eat their mates during copulation, beginning at the head so as not to interfere with the process. The perfectly good tasting viceroy butterfly avoids predation by mimicking the nasty monarch. Darwin's finches played a major role in his formulating and supporting his theory of evolution. And Karl Marx wrote to Darwin, asking permission to dedicate *Das Kapital* to the reclusive naturalist. That none of the preceding claims is true is unlikely to diminish the frequency with which they are told and retold. Dennis Chitty adds one more myth to the preceding list—that lemmings march off to the sea to commit mass suicide when their population numbers become too great.

Two complementary and sometimes antagonistic roles in science are the formulation of theoretically appealing hypotheses about how nature works and the testing of these hypotheses. Quite often one and the same person both formulates "beautiful hypotheses" and also tests them against "ugly facts." Dennis Chitty plays this dual role in the story that he tells about the science of ecology and evolutionary biology. Early in his career he formulated what he hoped would prove to be a "universal phenomenon," namely that "populations of all species living in favorable environments are capable of preventing unlimited increase in their numbers," and spent the next half century attempting to test his hypothesis by means of laboratory studies, field studies, and even what he terms "mail order zoology" from trappers, hunters, and game keepers.

By and large he failed. Each process that he thought might be responsible for cyclical fluctuations in the organisms under study, chiefly voles, did not pan out—starvation, shock, maternal stress, intra-group aggression, crowding—you name it. Nothing worked. Everyone knows about the difficulties of studying animals in the field, but laboratory studies are also replete with possible pitfalls. Too many things can go wrong, and they almost always do. For instance, Chitty spent some time trying to explain the mysterious, clumped deaths among voles being looked after by two of his colleagues, only to discover that the likely cause was water bottles becoming plugged up. Enlarged spleens in his laboratory voles also had a pedestrian explanation— fleas.

Chitty did not confine himself to testing his own pet hypothesis but also exposed the hypotheses suggested by other biologists to his searching criticism, including the golden cow of the day, density-dependent regulation. He complains that, unfortunately for ecology, the "theoretical hare hit the ground running long before the experimental tortoise emerged from the starting gate." Another unfortunate happenstance was that Wynne-Edwards' explanation of the regulation of population size in terms of group selection became the theory to refute. Refutations of group selection were read by many as refutations of the process as such, though Chitty's hypothesis garnered some press. Two of his papers became "Citation Classics," not because so many of his colleagues agreed with him but for the opposite reason. A negative citation is still a citation.

Like Darwin before him, Chitty is highly suspicious of the pronouncements of mathematical biologists. For example, he thinks that "mathematical modelers have done more harm than good through possessing a medieval faith in the power of logic to unravel secrets of the natural world." Chitty also passes some fairly harsh judgments on his fellow ecologists. In the absence of the experimental approach, he concludes that "ecology was doomed to remain glorified nature study," and characterizes the Wytham Ecological Survey as "19th-century natural history" and the term "scientific natural history" as an oxymoron. Needless to say, Chitty's facility with a stinging quip did not endear him to his targets.

Nearly all the books written about science (by scientists and students of science alike) tell success stories. Either a new synthesis was forged or the last piece was put in a very complex puzzle. For example, Darwin synthesized masses of data and apparently disparate phenomena in his overarching theory, and Watson and Crick were the first to come up with the molecular structure of DNA. Such episodes do occur in science, but they are the exception, not the rule. Very few scientists make a big splash. They put in the same amount of work, are equally dedicated and even "brilliant," etc., but the most that they do is to add a bit here and there to the edifice of science. However, when we tell nothing but success stories in science, most scientists simply drop out of the picture. Recounting the history of science entirely in terms of the rare scientist who really makes it big presents as inaccurate a portrayal of science as describing biological evolution only in terms of those species that have succeeded in surviving to the present or giving rise to other species that do. The vast majority of species go terminally extinct, only a few succeed in going extinct by evolving into later species, and even fewer make it intact to the present.

With respect to population cycles and density in ecology, Chitty rectifies this imbalance. No big breakthroughs have been forthcoming. One path is tried, then another. The problem situation is reformulated from the ground up, and still only small successes are realized. The story that Chitty tells is the typical story in science. However, because his tale is one primarily of failures interspersed with only an occasional advance, readers may well be put off. We are used to science marching on triumphantly. The glow of self-esteem reflects even on one's opponents. Winners can afford to be generous.

But what if everyone has failed? Then the criticism of one's opponents can appear to be ungenerous carping. Chitty comes down fairly hard on his opponents, especially such big guns as J.B.S. Haldane and David Lack, but he is equally hard on himself. He complains of the behavior of his opponents and then is forced to admit upon reflection that he himself has behaved in much the same way, although he could not see it at the time.

Over a half century ago, scientists began asking questions such as "Where do all the lemmings go?" Chitty is forced to conclude that we still do not know. He rejects the current received view in terms of density-dependent regulation, but he also has to admit that his pet hypothesis has hardly been established. But shouldn't every story have a conclusion? Maybe so, but insisting that every scientific episode must end in success seriously biases our understanding of science. If Chitty's book does nothing else, it shows that a life in science without a Nobel Prize is still worth living.

David L. Hull

Preface

In an honest search for knowledge you quite often have to abide
by ignorance for an indefinite period.

Schrödinger[1]

Once upon a time people believed that lemmings increased in numbers until
they ran out of food and then, in a recurrent 4-year cycle, took off for the
nearest cliff or ocean. Many scientists go along with the first part of this
explanation but doubt that lemmings are all in love with easeful death. Unfortunately, in spite of 70-odd years of learned effort there's still no answer
to this ecological equivalent to the riddle of the sphinx. Plausible hypotheses
do indeed abound, each championed by one of several quarrelsome cults; but
in spite of evidence in their favor, all lack the hallmark of scientific proof (so
called), namely, success in giving repeatable results. For until corroborated by
risky predictions,[2] favorable evidence must always be mistrusted, especially if
it's your own.[3]

This book is written for three kinds of readers: first, my successors in the
pursuit of an ecological Holy Grail, namely, the attempt to understand 4- to
10-year cycles in numbers of animals such as lemmings, voles, snowshoe
hares, game birds, and defoliating insects. I here describe why early ideas
failed to work, how others took their place, and why an experimental philosophy is needed to clear up the present confusion.

I also wish to interest scientists, whatever their specialty, who share my
fascination with the principles of scientific inference and their frequent di-
vorce from the way discoveries are actually made. Instead of the present title,
I might equally well have used the subtitle and called this book 'Beautiful
Hypotheses and Ugly Facts.'[4]

Third, the book is intended for anyone interested in mysteries, whether
scientific or otherwise,[5] for with one notable difference—the need to get repeatable results—it's like a detective story whose final chapters have yet to
be written. I've therefore tried to make the book readable by non-scientists,
who are naturally repelled by scientific jargon and gobbledygook. (Jargon is
the technical language that speeds up communication within professions—of
scientist, doctor, lawyer, psychotherapist, criminal—and slows it down between them. Gobbledygook is the ponderous prose and tortuous syntax due
to the use of abstract nouns, the passive voice,[6] and misplaced modifiers.)
Non-scientists will, I hope, see how much imagination, judgment, and conse-

quent error are associated with a profession whose objectivity they sometimes overestimate. Knowing the results of scientific research is less useful for them than understanding the type of mental activity involved in getting them. They may also see how much fun scientists get out of life and may lose some of their misconceptions about them.

"Truth will sooner come out from error than from confusion" wrote Francis Bacon;[7] but, sad to relate, most of one's time is spent in confusion, much in error, and little with truth. For there's no known way of proceeding except by a process of educated guess, trial, and error, especially error. And having to wait 4 to 10 years to replicate most observations is another reason it's taking so long to understand population cycles. As an early member of the group that began such studies, I've spent much of my career in confusion and error; but fortunately science is a self-correcting system that rises on stepping stones of its dead self to higher things. Or at least, it does so in the long run.[8] My own views about populations have changed several times already and will probably change again, so their present transitory form supports the need for more rigorous standards for testing all hypotheses, including the conventional view that numbers are regulated by density-dependent processes. Unfortunately for ecology, the theoretical hare hit the ground running long before the experimental tortoise emerged from the starting gate. This frequently means that modelers talk only to mathematicians and mathematicians talk only to God. Although theorists would have a hard time getting along without density-dependence, field workers should look critically at its track record and ask how much longer we must wait before it meets the test that "A good theory makes not only good predictions, but surprising predictions that then turn out to be true."[9]

The unifying theme of this book is the work on cycles at the Bureau of Animal Population, Oxford University during 1929–1961 together with a selected few of its later ramifications. The book provides a synoptic guide to this ancient literature but is no substitute for consulting it first-hand, as I have here reduced the evidence to its bare bones. One function of books like this is to provide a context that will encourage students to consult the primary sources.[10] Some of the diagrams are therefore given as the simple sketches drawn for anyone rash enough to ask about one's work or attend one's seminars. The original medium is likely to have been pencil on back of envelope, ball-point on paper napkin, chalk on blackboard, or felt pen on transparency. Such diagrams get to the root of the matter with less fuss than those that take account of raw and stubborn data. For 'evidence can be unreliable, and therefore you should use as little of it as you can.'[11]

This is not an account of science as it's supposed to be but as it was. Nor shall I try to persuade the reader that my conclusions are established by the data; I shall merely explain how I myself was persuaded by them. In due course it will be possible to see where my judgments went wrong; at the moment this is more obvious to my critics than to me. With the benefit of hindsight and the handicap of a selective memory, I may have got some of my wires crossed, but former students tell me the messages are not too badly scrambled. As some of these messages have been ignored, rejected, misinterpreted, or misquoted,[12] I explain the development of my views about the appropriate methodology and quote the authorities that have shaped them.

Ideas often go wrong, and seeing where they fail is at least as instructive as seeing where they succeed; they show man's infinite capacity for self-delusion, from which education, whether in science or the humanities, provides only limited protection. One can take the precaution, however, of recognizing that discovery is distinct from but dependent on justification. Their "methods and functions . . . are as different as are those of a detective and of a judge in a court of law."[13]

The cycle problem will be solved one day, perhaps when conclusions so far reached have been rejected in whole or in part and the decks are clear for new and imaginative solutions. A single paragraph in the text-books may condense the work of a century or more. But striking a balance between speculation and scholarship[14] is a perennial art to be learned and relearned in research of every description. I here present one such case history.

Department of Zoology,   D.C.
University of British Columbia,
Vancouver, B.C. V6T 1Z4
December 1995

Acknowledgments

The most important thing Monod learned from Morgan's group was the American style: the free, intensely critical discussions, the easy and open relations between colleagues of different ages and ranks.

Judson[1]

The custom of having lunch and afternoon tea in groups at the laboratory is a valuable one as it provides ample opportunities for . . . informal discussions.

Beveridge[2]

Here and there I leaven the text and notes with thumbnail sketches about the *dramatis personae*, of whom Charles Elton is the first and most important.

> *Charles Elton (Figure 0.1) began to study cycles early in the 1920s, and in 1932 established the Bureau of Animal Population as a separate unit within the Oxford University Department of Zoology and Comparative Anatomy, as it used to be called. He was Director from 1932 until he retired in 1967. He died on May 1, 1991 shortly after his 91st birthday. "In almost all subjects there are a few outstanding names. For it re-quires genius to create a subject as a distinct topic for thought."[3] Charles's name is indeed outstanding; but the following pages, which show how much I owe him professionally, are inadequate to show the affection with which I remember his kindness to Helen and me.*
>
> *At Charles's funeral, Harry Thompson (a wartime colleague) said those things all of us felt. He has kindly allowed me to include this eulogy.*
>
> > *Charles Elton's publications speak for themselves and for him.[4] After knowing him for nearly 50 years, respecting him and coming to love him, what can I add? That he was among the foremost pioneers of Animal Ecology is unquestioned. He has introduced generations of students to the importance of animal numbers, of cycles in numbers, and of animal communities in different habitats.*
> >
> > *The greater awareness today of man's profound influence on other species and the environment generally owes much to Charles's work; he urged constantly the importance of conserving nature in all its variety. Working quietly in his own way, eschewing committees whenever possi-*

Figure 1. In the beginning was Charles Elton, here seen in 1926 setting off with a load of mouse traps for Bagley Wood, the site of the first study of natural populations of mice and voles. Photo courtesy of Joy Elton.

ble, he nevertheless had a fundamental influence on Britain's official work of nature conservation. (As one of the senior staff of the Nature Conservancy said to me when Charles was giving one of his rare broadcast lectures: "our reponsibility is to conserve Charles Elton.")

His name is inseparable from the Bureau of Animal Population, which attracted zoologists from around the world, so that it became an honor to be numbered among Elton's Ecologists. In personal relations, Charles was as always, quiet, thoughtful, gentle, and persuasive, but firm in dealing with the problems inseparable from the role of Director.

He had great breadth of knowledge and an unfailing capacity for mak-

ing helpful suggestions when research projects ran into difficulties, as they always do. He read widely outside science and had a dry sense of humour. I am grateful for his having introduced me, long ago, to the works of such as James Thurber and Damon Runyon. At the end of your long and fruitful life, Charles, we salute you.

By leaving the Bureau in 1961 I was spared seeing Charles's unhappiness during his last years as Director. Judged from the Crowcroft account,[5] the abolition of the Bureau by the new head of the Department of Zoology was one of the more irresponsible actions of this home of lost causes. The high morale and enthusiasm for which, as a small, fiercely independent unit, the Bureau was world famous counted for little against the decision to merge it with the Department of Zoology. Administrators seldom realize the extent to which creativity need a special substrate on which to flourish and how easily it may be destroyed. But there was another side to the story, to which, from a safe distance, I was also sympathetic. Although I saw much of Niko Tinbergen[6] and his students, and although physiologists and geneticists were liberally distributed around the university, I had felt the need for the more frequent contacts one gets in a large department. This was one of my reasons for leaving to join the Department of Zoology at the University of British Columbia. Charles fought hard to preserve the Bureau as it was, and when he dug in his heels he was not the easiest man to deal with; nor from all accounts was the new head. So a clash of personalities was inevitable. I was glad not to be caught in the middle.

A free and easy style similar to that experienced by Monod was familiar to me from my 4 years with the Ontario Fisheries Research Laboratory. It was also Charles Elton's style, which accounts for much of the success of the Bureau of Animal Population, he himself having had no wish to be spared the criticisms of his colleagues. But the intellectual environment is only part of the picture. "We must have fun," wrote a Nobel Laureate,[7] "or our work is no good." Szent-Györgyi may have had in mind the work itself, but I like to think he included other types of fun (his nephew certainly did[8]). It was true of my time in Oxford that we not only worked hard but played hard, and that our days were full of excitement of all kinds. Those who play together stay together. Unlike people who have to punch a time-clock, we set our own gruelling hours of work.

No account of anyone's work is complete if it fails to recognize what it owes to other people. Luckily, I have always had colleagues who are critical but not too unkind and discouraging, as well as students from whose uninhibited comments I continue to profit. One of the advantages of going to scientific meetings is being able to associate a face and personality with work one has merely read about. The only substitute I can offer is to tell a few stories about my colleagues and former students. I need not have included all the stories, but together with Peter Crowcroft's account and photographs, they help put a human face on a process that may seem tedious and impersonal to those who took no part in it. They show that both fun and passion go along with academic work, and that a friendly social environment explains much of the success of an intellectual environment. It enables one to benefit from criticisms that would be hard to take from jealous or unfriendly colleagues. I am grateful to the many people, working on different problems and with different points of view, who have done their best to shape my education. I am also grateful to those who have read part or all of the early drafts of this book. Their names are as follows: Werner Baltensweiler, Katherine Biers, Charles Birch, Rudy Boonstra, Elizabeth Carefoot, Stephen Chittty, John Clarke, Peter Crowcroft, Joy Elton, Lee Gass, Marie Gibbs, Muriel Harris, Ray Hilborn, David Jenkins, Alice Kenny, Lloyd Keith, Charley Krebs, John Krebs, Xavier Lambin, Bill Lidicker, Peter Hochachka, David Hull, Sharon Kingsland, Helge Monsen, Judy Myers, Janet Newson, Rick Ostfeld, David Pimentel, Frank Pitelka, Dick Repasky, Don Reid, Tarald Seldal, Kitty Southern, Carole Stanley, Nils Stenseth, Bob Tamarin, Harry Thompson, Andrew Trites, Leila Vennowitz, Adam Watson, and Jean Wilson. I am especially indebted to my wife, Sherry Kendall, for having restored my peace of mind and in many other ways enabled me to write this book. Finally, I'm grateful to Andrew Trites for the many hours he spent dispelling my computer illiteracy; also to Rudy Boonstra, who came to my rescue on several occasions.

The work in England was supported by a variety of organizations.[9] The work in Canada was supported by the National Research Council and later by the National Science and Engineering Research Council. The latter blotted its copy book by confiscating a few unspent dollars I had saved to support work after I retired. Before the end of the financial year I should have claimed for fictitious miles of travel or squirreled the money away into another account. In one's dealing with bureaucrats, honesty is seldom the best policy.

Contents

Introduction

We have a habit in writing articles published in scientific journals to make the work as finished as possible, to cover up all the tracks, to not worry about the blind alleys or describe how you had the wrong idea first, and so on. So there isn't any place to publish, in a dignified manner, what you actually did in order to get to do the work.

Feynman[1]

Only rarely does someone let down the hair and baldly relate things as they really happened.

Vogel[2]

1.1 Preview

On a visit to Norway in 1923, an Oxford undergraduate, Charles Elton, became intrigued by stories about the apparently suicidal migrations of lemmings. Six years later he began a series of studies in which, 6 years later still, I was lucky enough to join. Now, more than half a century later, I'm still trying to separate fact from fiction, much of the latter created by myself in pursuit of the former in the perennial scientific conflict between imagination and "irreducible and stubborn facts".[3] In the account that follows I try to describe this conflict chronologically; but because many things were going on at the same time, because topics such as food, for example, were studied intermittently over the years 1929-1961, and because I need to relate these topics to some that followed, no sequence can be entirely satisfactory.[4] To add to these difficulties, much of the work was unpublished until many years after it had been finished, and some was not even analyzed. Novelists have their own chronological problems and solve them with such phrases as "meanwhile back on the farm . . ." I shall play similar tricks with a sequence that goes somewhat as follows.

After returning from Norway, Charles Elton had the problem of deciding how someone living in Britain and working on a financial shoestring was to find an experimental animal through which to discover principles that might apply to lemmings in Norway and perhaps to other species elsewhere. He chose the short-tailed field vole, a small mouse-like creature that became the equivalent for some population ecologists of the geneticists' fruit fly. Chapter 1 introduces the lemming problem, discusses the need to select the right questions, and explains the risk of making the wrong assumptions; for populations of voles are not necessarily analogous to populations of lemmings and other animals, and mere case histories are of limited use unless they can be used for or against some general idea.[5] But the vole was worth taking a chance on, as the cost of being wrong was less than the possible benefit of making new discoveries, testing the generality of one's ideas, and reducing the need to reinvent the ecological wheel for each species. People differ, however, in deciding what assumptions are justified and which observations are analogous. Such differences are the salt that seasons scientific speech, not always with grace, for scientists are no more immune than other mortals to getting hot under the collar. The chapter ends with a preview of the Chitty Hypothesis, in which I naturally have a vested interest.

Chapters 2 and 3 describe the first 10 years of work on the vole, show the eventual bankruptcy of the simple ideas with which it began, and introduce the idea (new at the time)[6] that changes in the *kinds* of animals go along with changes in *number,* and may help explain cyclic declines. These ideas were put on hold during World War II, when we applied our ecological skills to killing rats and mice (Chapter 4) as well as rabbits. The first postwar job was to replicate the prewar observations and define the components of a population cycle (Chapter 5). Given only vague descriptions—for example, merely that animals have disappeared over winter—authors can get away with almost any explanation: bad weather, predation, starvation, disease, dispersal. Without knowing whether age- and sex-specific survival rates are uniform or variable, no one can test the author's interpretations. A prerequisite, therefore, to deciding between rival explanations is to have accurate and frequent natural history observations; many a bright idea will then fall on stony ground.

Chapter 6 explains how we began testing implications of the idea that populations change in quality, and Chapter 7 describes the chastening opposition this ran into. Chapter 8 describes innovative work on partridges that resulted from trying to explain cycles through studying behavior instead of following conventional wisdom and studying food, predators, parasites, and disease. In voles, this alternative approach, though fruitful, was still insufficient to explain the difference between populations that made rapid gains one year and, under apparently favorable conditions and at similar population densities, suffered catastrophic losses a year or so later. I now thought the difference in outcome was associated with a difference in selection pressure, having reached this view through serendipity rather than logic. The chapter ends by suggesting that vole cycles are a special case of a universal phenomenon, namely, that populations of all species living in favorable habitats are capable of preventing unlimited increase in their own numbers. Not all populations actually do so, however: those in unfavorable habitats may have no surplus to get rid of; predators may do it for them.

Chapter 9 describes further tests of these ideas, and shows that lemmings in the Canadian arctic do indeed behave analogously to voles in Great Britain. Chapter 10 presents evidence that among populations of lynx, muskrats, game birds, and snowshoe hares, some populations fluctuate in step, but that hare and game bird populations in eastern Canada may be totally out of phase with those in the west. A model explores the idea that weather is the regional synchronizing agent. Chapter 11 reviews the growth of my own ideas about the regulation of population density and appropriate scientific methods. Chapter 12, the epilogue, explains why some critics consider my methods misguided, touches lightly on later attempts to understand vole and lemming cycles, and suggests which lines of enquiry are worth pursuing.

In research as in ordinary life the truth is rarely pure, and never simple, but the present unvarnished account may enable other people to see whether it's purer and simpler than I think it is. The impeccable authorities I quote throughout this book have been of little help in discovery, but of immense help in disciplining my thoughts, or so I like to think. Some authors maintain that if classical criteria of evidence are too demanding, so much the worse for classical criteria.[7] This attitude offers a convenient excuse for lack of critical thought. McIntosh, for example, states that "Ecology was not and is not a predictive science."[8] If projected into the future, this judgment puts ecology forever beyond the scientific pale, where, according to unkind colleagues in physiology and genetics, it resides at present. Their misgivings are unlikely to be banished in the following pages.

As most of us used first names, I find it unnatural to refer to old friends by surname only. The first person plural refers to different colleagues at different times. The style is autobiographical, the rationale being as follows:[9]

> To grasp a scientific idea it is, of course, not nearly enough just to have a bare statement of it. One has to appreciate the programme behind it (e.g. the unification of two separate sciences), see what demands it satisfies, what differences it makes. In short, one has to understand its pre-history and contemporary context.
>
> Usually, the person with the best knowledge of an idea's background is its author; and if he gives a revealing account of this, it will inevitably have an autobiographical flavour. He will explain how *he* came to recognize a certain problem (for problems are like pieces of furniture, man-made but public). He will describe *his* original approach to it and, perhaps earlier attempts to solve it which proved abortive but which helped him to understand it better.
>
> Consequently, a scientist who supposes that it is somehow *de rigueur* to present his ideas in a non-autobiographical and impersonal way, can hardly provide the background needed for an adequate grasp of them. . . .
>
> Scientific discoveries are remarkable things and we are entitled to the truth about them. It would be better if their discoverers told the full inside story in the first place, instead of presenting it circumspectly, as if there were a mysterious 'fifty-year rule' which obliges them to leave it to the historian to unearth what really happened. For apart from the delay there is the real danger that the historian will get it wrong anyway.[10]

To the sceptic who says "Who cares anyway?" the short (and fallacious) answer is that pure research usually pays off in hard, long-term currency. If we wish to control outbreaks of locusts, defoliating insects, or rats and mice; if we wish to conserve endangered species, manage grizzly bears and other animals in the national parks; if we wish to help politicians save our fisheries from collapse—if we wish to perform these and other good works, we must first understand the principles of population ecology. Unexpectedly, as is usually the case, the present research paid off in applications to medicine, pest control, and conservation.

A better answer to the sceptic is that the main educational function of universities is to help young men and women develop imaginative and critical faculties they can apply in the outside world. Science departments can usually discharge this function better and at less cost to the taxpayer by putting students to work on pure rather than applied problems: on the behavior of beetles rather than bears. But the real answers are more complicated, as another function of universities is to provide a safe haven for people with novel and sometimes unpopular ideas. Such people are driven by mental states that range from a belief that wisdom is more to be valued than gold, yea than much fine gold, through a wish for fame to the "wow-feeling of discovery. . . . Like orgasm, it is something [. . . to be experienced] as often as possible."[11] Fortunately for such people, advanced societies think it worth supporting at least a few of them, especially as the titles of their articles are always good for a laugh.[12]

The main technical issue in the present story is whether the key to understanding populations of lemmings and other species is to study the behavior, physiology, and genetics of the species or to continue restricting one's study to its food, cover, enemies, and disease. Making the wrong choices can be costly in time, money, reputation, and printer's ink; so a fundamental issue is the need for population ecologists to agree on standards of evidence. It's not enough to say "When in doubt appeal to experiment;" experiments can be misinterpreted and are difficult to replicate; one must also appeal to prediction, which again is difficult, as will soon become apparent; and one must also appeal to theory, though none but natural selection can be relied on at present. So, for the time being, it's open season on everybody's ideas, and authors should be gratified rather than distressed to find that their ideas are considered worthy targets.

Writers of scientific papers often take the following advice: (1) say what you're going to say, (2) say it, (3) say what you've just said. Readers can then skip the details of methods, materials, and results (step 2) until they have digested the introduction (step 1) and the discussion and conclusions (step 3). They can then go back and examine the evidence. This is the way I find it best to cope with scientific papers and recommend it to readers of the present book. Chapter 1 corresponds to step 1 and Chapter 11 to step 3. The intervening chapters, epilogue, and notes contain supporting details for the benefit of my colleagues and of some interest, I hope, to other readers. They add further details to a typical story of how much time-consuming and apparently fruitless research may be needed to answer even a simple question such as "If lemmings don't commit suicide, just what do they do?" I wish I knew; for the best answer I can give—that selective advantage switches back and forth between two kinds of behavior—is based on assumptions about the accuracy of a number of unreplicated observations, on arguable interpretations of data, on supposed parallels between diverse organisms, on the assumption that cycles are sufficiently alike to justify a search for common antecedents, and on the philosophy that rival ideas are no threat until they have been tested in well-replicated experiments with adequate controls. It's obvious, therefore, that my chain of reasoning has a number of links that need to be strengthened or replaced. It's up to students to spot them. A good exercise would be for students in successive classes to predict *in writing* how much, if any, of this and rival schemes will stand the test of time. With luck, the accuracy of their predictions will be revealed some time in the course of their professional careers.

1.2 Lemming stories

All—or very nearly all—scientific theories originate from myths.

Popper[13]

O that the Everlasting had not fix'd
His canon 'gainst self-slaughter.

Hamlet

In 1923, Charles Elton returned from Norway with stories about the spectacular rise in numbers of lemmings, their sudden and regular disappearance from the mountains, their long-range movements, and appearance in unusual places. Some of these stories should now be taken with a grain of salt; this one, for example:

> It is astonishing that so many reach the coast alive. Yet in 1868 so many swam out into the inner parts of Trondheim Fjord that a steamer took a quarter of an hour to pass through them.[14]

These so-called migrations are often mere travelers' tales. One need not doubt that the land is sometimes overrun with moving animals (Figure 1.1), but without marking experiments we can seldom say how many are milling around rather than moving through.[15] Large movements have yet to be filmed by reputable observers, and I know of only one, rather blurred photograph of a lemming in transit;[16] it's hardly the stuff of legend (Figure 1.2). (The stampede of lemmings in Walt Disney's *White Wilderness* is an obvious fake.[17]) In atypical habitats, lemmings are indeed noticed in peak years only; but even in typical habitats few or none are caught between peaks. At Point Barrow, Alaska, for example, no collared lemmings were caught in 7 of the 15 years preceding their resurgence in 1970 and 1971 and only one to five were trapped in other years (Section 12.3). In Scandinavia there seems to have been no long-term trapping in marginal habitats to show whether lemmings invade them or, as in good habitats, are present between peaks but indetectable. Authors must refute this possibility empirically rather than verbally if they wish to dispel the recognized North American scepticism[18] about the reality of journeys of 100 km or more. Shorter movements, of course, are entirely credible and seem to be similar in scale among Old and New World lemmings. More must be done to separate myth from reality.

Reports of movements in the Canadian arctic turned up during an enquiry[19] carried out between 1935 and 1949 (Figure 1.5). Here is an unedited sample of how things looked to our correspondents:

> **Read Island** [in the SW corner of Victoria Island]: In the fall of 1938 lemming were everywhere. This spring [1939] there is not one to be seen. Where they went is a mystery. No dead ones have been found. In 1941-42 lemming were observed on sea ice nearly every month during winter and then again during this month, May, but it is difficult to say whether they were migrating as all their tracks ran in circles. Most of the lemming and their tracks were seen close to shore—about 50 yards or so off the beach, but in May lemmings were seen ten miles off shore on salt ice.

> **Coppermine.** [160 km S of Read Island]. Very large numbers of lemming noted far out on ice where many died April 1945.

> **Chesterfield.** Large numbers seen on the sea ice near floe edge in May [1945].

These reports are more objective than most; the following mix observation and interpretation, a fault to which scientists are also prone, as will become apparent. The reports nevertheless confirm the suddenness of changes in abundance and extent of movement onto the sea ice.

> **Tavane.** During April and early May [1940] exceptional migration of lemming observed and reported by various sources from Eskimo Point to Chesterfield Inlet. These animals were apparently travelling eastwards, and large numbers (several thousands at least) were seen perished on the salt water ice.

> **Coppermine.** Never saw so many lemmings before; they migrated here 20–25 September [1940].

> **Baker Lake.** Lemming were abundant in the summer of 1943. In November they migrated, presumably toward Eskimo Point. In May 1944 they were practically non-existent. Numerous carcasses were seen lying in the glare ice on the lakes as if they had frozen to death,—none seen on land, no disease—(reports indicate they migrated towards the timber line).

> **Pangnirtung.** "March of Lemming" first started around Kivitoo and Broughton Island in early February [1945] and spread southwards, reaching Cape Mercy and mouth of Cumberland Sound by April [about 300 km SSE of Pangnirtung].

Lemmings that move onto the sea ice will almost certainly die from lack of food and shelter, and in a sense are committing suicide. So it's easy to see how the myth arose. But it's harder to believe that suicide is what's on a lemming's mind. For one thing, suicidal lemmings would leave fewer descendants than those that stayed home, and natural selection would put an end to so simple a solution to overpopulation (Figure 1.1). As in other species, however, emigration enables some animals to find sanctuary away from an overcrowded habitat.

THE FAR SIDE   By GARY LARSON

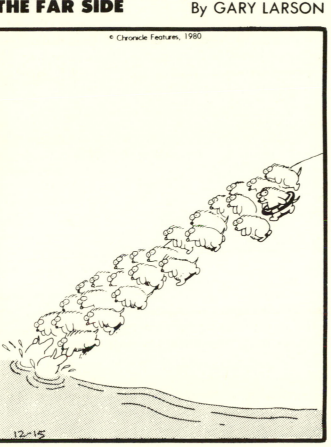

Figure 1.1. Lemming migration: as expected. Natural selection would favor descendants from the aberrant animal at the rear. *The Far Side* cartoon by Gary Larson is reprinted by permission of Chronicle Features, San Francisco, CA. All rights reserved.

When lemmings become scarce at one place and abundant later on at a place some miles away, the two events seem to confirm the animal's reputation for long-distance running. But the timing could be coincidental. Lemmings on the sea ice have obviously 'migrated', and long-range movements on land would explain the sudden absence of lemmings as well as their apparent arrival from elsewhere. But as one sceptic said of their disappearance, "where they went is a mystery." He put it better than he knew; 57 years later we still don't know.

According to some observers, lemmings were more abundant after the winter than before, a belief that could have been merely apparent, owing to the disappearance of snow and increased activity in spring. If real, the increase could have been due to immigration or, as some observers suggested, to breeding; but as winter breeding was contrary to everything known about

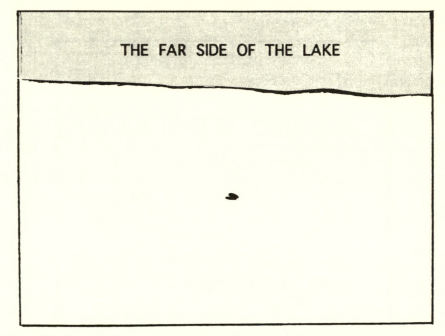

Figure 1.2. Lemming migration: as observed. Drawn from Koponen et al. (1961, Fig. 3): "A typical scene during the migration: a single lemming [and its shadow] crossing the ice."

arctic breeding seasons we dismissed this as idle speculation. We changed our minds later.[20]

These stories contribute little to solving the lemming mystery, but remind us that it fascinates normal people as strongly as those who pursue the problem with the obsessive zeal of captain Ahab in pursuit of the great white whale. The interest of the general public is shown in passages such as the following:

> On any day when the wind is up, you only have to raise your eyes to watch the clouds hurling themselves like lemmings over the rim of Table Mountain.(Peter Taylor in the *Independent* Magazine, July 18, 1992.)

> Like lemmings joyfully jumping off a cliff, Conservatives ate their leader Friday night . . . (Aileen McCabe in *The Province*, Vancouver, January 30, 1983.)

> Intervention by the world's central banks failed to halt the dollar's lemming-like dash for the cliff edge. (*Manchester Guardian Weekly*, August 30, 1992.)

> Bankers make lemmings look like free-thinkers. (Quoted by Geoffrey Rowan in *The Globe and Mail*, Toronto, December 26, 1992.)

> Names are leaving [Lloyd's] not like sheep but like lemmings. (Julian Barnes in *The New Yorker*, September 1993.)

> Hot women flocked to the Come Clean Center like lemmings headed for the Sea of Matrimony. *(Tales of the City* by Armistead Maupin.)

It would be a pity if our cultural gene pool were impoverished by doubts about the validity of such similes. But science marches on and has already demolished some colorful explanations. Here is one attributed to the Laplanders:

> The lemming is the reindeer's little brother. We know this because it has a short tail, it lives in the mountains and it causes no harm.
>
> But where do all the lemmings come from?
>
> Far, far away among the mountains there is a living fell. It is so tall that it towers high above the clouds and no living creature can ever reach the top. This mountain gives birth to the lemmings. They stream forth, never stopping. There are so many of them that they crawl over one another. If a man could climb up there he would not be able to stand it.
>
> In some years the lemming-fell becomes so full that it can not give birth to any more lemmings. Then the fell calls on the clouds, wind and rain and begs them to carry away as many lemmings as they can and to drop them down here and there all over the country.
>
> This is why hoards of lemmings sometimes fall from the sky. Many of them die then. The rest gather together in large flocks and run from the place in search of the mountain which gave birth to them. But they never reach it.[21]

As a temporary substitute for the truth, a hypothesis is useful to the extent to which it is fruitful, that is, promotes enquiry. Myths once served this purpose; some still do, for any hypothesis is better than none, provided it is looked at critically.

1.3 Lemming cycles

And such collection must be made in the manner of a history,
without premature speculation, or any great amount of subtlety.
                                                    Bacon[22]

Stories in the last section were incidental to the main purpose of the Canadian Arctic Wild Life Enquiry, which asked whether lemmings and their predators were more, less, or equally abundant compared with the year before. Where the contrast was pronounced, replies were probably accurate.

By analogy with Norway, where fox and lemming cycles were known to be associated, it seemed likely that a cycle in lemmings might also be linked to the cycle in foxes in Canada.[23] Figure 1.3a shows two 4-year cycles in the number of fox pelts sold by the Hudson's Bay Company.[24] From peaks of about 5000 in 1901 and 1905, sales dropped to about 200 two years later. Trappers caught in this boom-and-bust economy suffer in lean years from having few skins to trade for goods at the company stores. The harsh reality for them is the *change in number* of foxes. The population biologist, however, is more concerned with the *change in rate*, regardless of population size. Such changes are more easily compared if a drop of (say) 50% is represented the same way whether the loss is 50% of 10,000 or 50% of 100. The trick

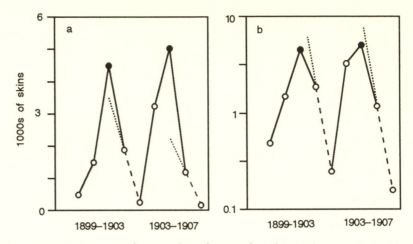

Figure 1.3. Two 4-year cycles in sales of arctic fox skins (Elton 1942: Table 51), plotted so that the slopes of the lines show (a) changes in numbers and (b) changes in rates.

to making such comparisons is to plot data in logarithms (Figure 1.3b), in which case the slope of the lines shows whether rates of loss (or gain) are the same or different.

One obvious difference between the two ways of plotting the present data is shown in the second year of the declines. In Figure 1.3a the slopes, plotted arithmetically, are less steep in the second year than the first; those in Figure 1.3b, plotted logarithmically, are steeper. Although Figure 1.3a shows that the *number* of pelts traded was lower in the final than in the previous year, Figure 1.3b shows that the *rate of decline* was higher. Higher rates of decline in spite of declining numbers are characteristic of cycles and have still to be explained. Because we need to study such rates of change, population data in this book are plotted in logarithms. Too many ecologists continue to use arithmetic plots; at one time I myself knew no better.[25]

Figure 1.4 shows that the sale of pelts of the arctic fox had a regular 4-year cycle for at least the quarter century between 1897 and 1921 and a less regular fluctuation in earlier years. These fluctuations are presumably due to fluctuations in the number of foxes, modified at times by irregularities in transport, marketing, or record keeping. And fluctuations in the number of foxes are presumably due to fluctuations in the number of lemmings (and not vice versa, as some theories suggest).[26] Peaks in the collection of pelts do not, however, necessarily coincide with peaks in lemming populations, nor should sales be mistaken for population data. For one thing, foxes are less likely to be caught when they have plenty of food than when they are hungry, which will usually be in the year after their prey has become scarce. Moreover, foxes are trapped over large areas and several months, so the resulting statistics cannot always be related to local fluctuations in numbers of lemmings.

Snowy owls provide further indirect evidence for the lemming cycle. When

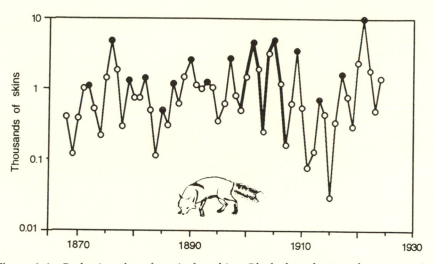

Figure 1.4. Cycles in sales of arctic fox skins. Black dots show peak years as given by Elton (1942: Table 51). Thick lines show the data plotted in Figure 1.3.

their prey is abundant, snowy owls raise large broods, which results later on in too many owls competing for too few lemmings. In searching for alternative prey, many owls fly into populated areas far south of their breeding range. Records of these invasions have also had a periodicity of 3 to 4 years.[27]

In 1935-1936, the first year of the enquiry, lemmings were said to be scarce almost everywhere in the Arctic. In the next year, reports of recovery were common throughout the Eastern Arctic, and a peak was expected in 1938-1939, 4 years after the last one. It was therefore with high hopes of seeing the land overrun with lemmings that in July 1939 I joined the Eastern Arctic Patrol[28] aboard the R.M.S. *Nascopie* at Churchill, Manitoba. Our route took us up the west coast of Hudson Bay, up the east coast of Baffin Island, north to Ellesmere Island, and west to Fort Ross at the entrance to the Northwest Passage. On the final leg of the journey we came down the Labrador coast, finishing up at Halifax on September 12, 1939.

The main biological reward for 7 weeks and thousands of miles of travel was seeing masses of ancient lemming droppings.[29] Without this evidence, one might have concluded that these coastal areas were outside the lemmings' range. At Chesterfield Inlet and Fort Ross, however, I was able to catch a few animals by hand and carried them to Toronto, where I abandoned them. For war had broken out, and studying cycles had become a trivial pursuit. One other reward, disproportionately small for the length of the trip, was personal confirmation that correspondents could tell a low year from a peak. For when we analyzed replies to the questionnaires, we found all but 8 of 36 observers in the Eastern Arctic (excluding Labrador) had reported decrease or scarcity among lemmings in 1938-1939. Four of the eight reports were from west of Hudson Bay, where the decline was absent or less marked than in northern Quebec and Baffin Island.

By bad luck I had struck a 3- instead of a 4-year cycle, and 50 years later, in September 1989, I suffered a similar disappointment by arriving at Finse, Norway, 6 months after lemmings had declined from an unusually high peak. On both occasions it was hard to believe any lemmings had survived and easy to appreciate early myths that the land is recolonized from the sky and modern myths that some areas are recolonized by long-distance migration.

Inability to predict population trends or, given a new population, to tell what phase it's at, continues to upset the best-laid schemes. As a result, I was given a hard time by some of the practical men aboard the *Nascopie*, Charles Elton having made an overconfident prediction that 1938-1939 would be a peak year. This was an excusable mistake in view of the primitive state of the art, which included a belief that a 4-year interval was almost a law of nature. Mathematical precision, however, is not an essential characteristic of these biological cycles, so I shall continue to call them cycles on the assumption that they are imperfect copies of some Platonic ideal. Purists prefer the cumbersome term 'multiannual population fluctuations' and draw artificial distinctions between those that do and those that do not conform to their preconceived ideas about the significance of regularity.[30]

Even if I'd seen lemmings everywhere I could have done little more than begin educating myself about their habitats. And with the outbreak of war further work was impossible. Luckily, Helen was able to keep the enquiry going until 1949, by which time it had revealed as much as it usefully could. The enquiry was not designed to solve the lemming problem but to describe it in more detail than was otherwise possible at the time. The best attempt to do so had been made by Shelford,[31] who presented curves for 1929-1943 based on "a rough expression of experience with the lemming rather than an accurate year to year determination by sampling, but based primarily on trapping." Although his samples were too few and his techniques too primitive to establish peak densities or rates of change, he produced unusual evidence about the extreme scarcity so typical of the low phase. In July 1939 (the month I passed through Churchill) he walked 1 to 4 miles between fresh signs, which was an improvement over 1931 and 1934, when he found no fresh signs at all. And for 1932 he believed there were no more than two lemmings per hectare—despite an estimated 30-fold rate of increase, which his curves, plotted arithmetically, fail to show. Not until a quarter century later was precise work done on populations of lemmings in the Canadian Arctic.

Figure 1.5 gives an example of what an enquiry does best, namely, shows which populations are going up or down at the same time. Such information is difficult and expensive to obtain through personal visits. The map shows that by 1940-1941 lemmings were recovering from the scarcity I'd witnessed in 1939. They then decreased in one or both following years, at least in northern Quebec and on Baffin Island, and increased again in 1943-1945. As a result, the map for 1944-1945 resembled that for 1940-1941, and 2 years later the map for 1946-1947 resembled that for 1942-1943. Thus, in this part of the world, comprising thousands of square miles, lemming popula-

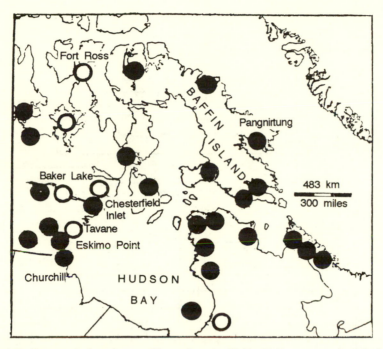

Figure 1.5. Reported abundance of lemmings in 1940–1941 compared with the year before. ● = increase, ○ = decrease. See H. Chitty (1950b).

tions kept more or less in step; but to the west of the area in Figure 1.5 fluctuations were less obvious or in a different phase, especially in 1942-1943. These regional differences helped kill the original idea that there was a circumpolar synchrony in lemming cycles. Nor at any one place was the cycle as regular as the cycle in fox pelts, though departures from expectation could have been due to faulty observation. As far as one could tell, there were no fundamental differences in behavior between Old and New World populations of lemmings other than those imposed by differences in topography.

The enquiry confirmed that flights of snowy owls were associated with changes in the numbers of their prey. Arrivals in the south were observed in 1934-1935, 1937-1938, 1941-1942, and 1945-1946, 1 year earlier than reports of general scarcity in lemmings.

1.4 Relevance

The separation of relevant from irrelevant factors is the beginning of knowledge.

Reichenbach[32]

Figure 1.6 shows some of the factors likely to affect lemming populations.[33] Making diagrams of this kind is a popular sport among ecologists and con-

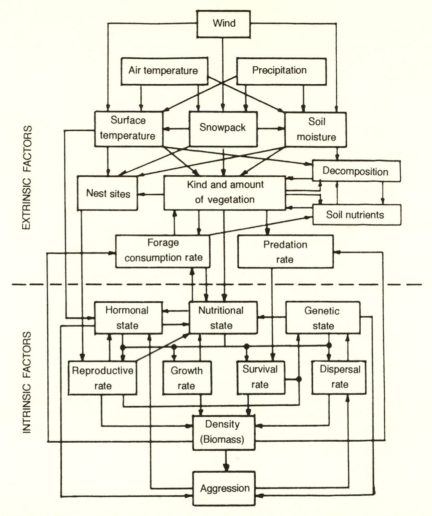

Figure 1.6. Factors influencing the population density of lemmings. From Batzli et al. (1980).

sists of drawing arrows between boxes to indicate interactions between variables. There are already 58 arrows in Figure 1.6, and more might have been added if the authors had included bacterial and viral infections, endo- and ectoparasites, competitors, infanticide, and a range of local peculiarities that make every population unique. With so many interacting variables, it would be unrealistic to study them all at once; so one begins by selecting some aspect of the whole, restricting one's research to those that can be studied by precise, repeatable, and preferably quantitative experiments. Some workers nevertheless object to studying a complex biological problem by first chopping it up into small pieces. For if everything affects everything else we shall be unable to isolate one fragment without destroying our understanding of the whole. The best defense for studying parts of the problem in isolation is that the method seems to work in the end, that with limited resources we have no alternative, and that progress has been made in other fields by selecting specific and limited problems instead of trying to explain everything at once. Some naturalists are so impressed with the complexity of nature that they find it hard to understand that progress depends on scientists becoming "content with little questions that can be answered."[34]

Having chosen a phenomenon to study, we next have to discover the relevant (or necessary) factors. A relevant factor is one that must be mentioned in any explanation of the selected phenomenon;[35] irrelevant factors are those that may modify it but are not common to all instances. Factors that are irrelevant to one aspect may be relevant to another, and we expect people with other interests to study them so that some day we can pool our knowledge. Although "our scientific conception of the universe . . . is made up of abstractions" we must avoid the error of "misplaced concreteness, namely, of mistaking abstractions for concrete realities.[36] They should be regarded as an intermediate stage between ignorance and an ultimate but perhaps unreachable holistic explanation.

Instead, therefore, of trying to explain a phenomenon in its entirety, one begins by trying to explain *the difference between it and its control.* Making comparative statements is one of the tricks of the scientific trade. Indeed, without controls—or the nearest ecological equivalent—one is left with an unmanageable set of variables,[37] which is the fate of many population studies. The control for a regulated population is an unregulated population—one that is increasing exponentially—and we understand the former in terms of its difference from the latter. With a recurrent phenomenon such as a cyclic decline in numbers we therefore look for recurrent factors, and eliminate as irrelevant those common both to the decline and its control. (Factors so eliminated are, of course, irrelevant only to the *difference between the phenomenon and its control,* not to the phenomenon itself.)[38] Eliminating irrelevant factors from one's list does not, unfortunately, guarantee that it will now include the remaining relevant factors; it will include them if and only if they have been thought of in the first place—a difficulty overlooked both by Sir Francis Bacon and by Sherlock Holmes. In a new field, the main function of elimination is to get rid of early preconceptions and show the need to invent alternatives. These in turn will be subjected to tests of relevance in a recurrent cycle of their own.

It's trivially true that the sum total of all interacting factors is *sufficient* to produce any natural phenomenon, and that in retrospect one can explain it to one's own satisfaction by putting in enough of these factors. But the more factors one puts in, the more replicates are needed to test the model,[39] which in population ecology means that it will be untestable. Moreover, if the premises are incomplete or false, the conclusions may follow logically but are unlikely to be true; at best they may be right for the wrong reasons. Modelers who are not also field workers are notorious for confusing validity with material truth, and field workers who are not also logicians sometimes mistake pure reason for scientific evidence. Although chided for failing to incorporate my own ideas in a mathematical model,[40] I believe that until we know the relevant factors this approach to knowledge is premature and, except when used as an elaborate guide to discovery,[41] may lend a false sense of security or be downright misleading. My respect for the intellect of many modelers is tempered by a Baconian "impatience with builders of whole systems out of scant materials."[42] The reader should, however, consult Sharon Kingsland's *Modeling Nature*[43] to get a balanced view of the tensions between those who prefer order to the chaotic observations of the field worker and those who wish theorists would keep in touch with the real world. That *Modeling Nature* is well balanced is clear from complaints from some mathematicians that the author is too critical of them and from some empiricists that she lets them off too lightly. Hansson and Stenseth[44] provide examples of models supposed to improve our understanding of cycles in small rodents. The strong natural history tradition in ecology and lack of an experimental philosophy may explain the authors' preference for verbal and mathematical descriptions to less ambitious comparative models.

1.5 Ideas Right and Wrong

An art of discovery is not possible; we can give no rules for the pursuit of truth which shall be universally and peremptorily applicable.

Whewell[45]

There is only one proved method of assisting the advancement of pure science—that of picking men of genius, backing them heavily, and leaving them to direct themselves.

Conant[46]

Changing myth into science is less easy than it sounds, though not all problems take as long to solve as it's taking to explain cycles. Science is not an objective skill of applying something called "scientific method," but a fallible creative art. Just how fallible will become obvious from the present book. Most research includes a preliminary "state of imaginative muddled suspense."[47] This state of mind is normally concealed, and "the scientific paper in its orthodox form does embody a totally mistaken conception, even a travesty of the nature of scientific thought."[48] Dubos discusses this issue as follows:

> Anyone interested in the performance of the human mind—out of mere curiosity, or for scholarly pursuit—welcomes the publication of the tentative and often crude sketches through which artists and writers evolve the final expressions of their ideals. However, scientific workers now consider it unbecoming and compromising to reveal their gropings towards truth, the blundering way in which most of them, if not all, reach the tentative goal of their efforts. This modesty, or conceit, robs scientific operations of much human interest, prevents an adequate appreciation by the public of the relative character of scientific

truth, and renders more difficult the elucidation of the mechanisms of discovery, by placing exclusive emphasis on the use of logic at the expense of creative imagination. The raw materials out of which science is made are not only the observations, experiments and calculations of scientists, but also their urges, dreams and follies.[49]

In the orthodox form of a scientific article, conclusions emerge logically from objective data instead of chronologically from one faulty hypothesis after another, from accidents, leaps of the imagination, and risky judgments, and from the constructive and destructive comments of colleagues, students, editors, referees, and readers of the scientific literature. Most of the work in the present book has appeared in orthodox form; here I describe the tortuous paths that led to the stage reached by 1961, after which I bowed temporarily out of the race. All conventional explanations for cycles had, so I thought, been discredited some 20 years earlier, and "no selection, no cycle" was the idea I hoped my successors would test. A few did so, though less persistently than I had expected, and most continued to look for explanations based on the discredited method of correlation rather than experimentation. If these workers had nevertheless solved the problem, they would have saved me the trouble of writing the present book; but as they have not done so, they may learn something from my mistakes.

Attractive ideas are easy to come by in population ecology but hard to test. For although the principles of testing are explained in many a text-book, applying them to population data is something else again. I describe my own attempts to do so in the hope that even if wrong about cycles, I'm right about the need to apply to ecology the same experimental reasoning that is axiomatic in other biological sciences. Long-term studies, the sacred cow of some ecologists,[50] are of limited explanatory value unless combined with experimental work, and authors who dispense with controls must be correspondingly modest about their conclusions. Indeed, "unless the basic needs of the controlled experiment can be satisfied it is better to abandon the attempt [to do experiments]."[51] Uncontrolled descriptive studies are even less trustworthy. Controls are particularly necessary in applied ecology because of the inexplicable ways in which populations rise and fall or switch to new levels of abundance.

Ecologists who believe they can prove their hypotheses are not alone in their delusion. Even physicists sometimes believe they can do so. Newton huffed and puffed when some incautious 'Rev. Father' dubbed one of his views an ingenious hypothesis—"Scientists did not speculate about things they could not prove."[52] We no longer think this way. In fact, the best we can do is to keep working on an idea, however imperfect, until something better comes along; there is no certainty about scientific proof, so-called. Huyghens[53] recognized the limits to the power of science when he wrote:

There is to be found here [in natural science] a kind of demonstration which does not produce a certainty as great as that of geometry and is, indeed, very different from that used by geometers, since they prove their propositions by certain and incontestable principles, whereas here principles are tested by the consequences derived from them. The nature of the subject permits no other treatment.

Another way of making the same point is to say that deductive conclusions (from general to particular) follow necessarily from the premises, whereas scientific or inductive conclusions (from particular to general) are at best only probable, so must be judged by prediction. Consistent success will justify but not prove one's conclusions, and for a different reason (see item 6 below and section 2.9) consistent failure may discredit them but will not prove them wrong. Proof and disproof in science are thus equally impossible.

We can summarize sections 1.5 and 1.6 as follows. (1) Anyone with ingenuity can explain events after they have happened, but (2) such explanations, however pleasing to the imagination, are useless if they merely 'predict' the data from which they have been derived; this is a form of circular reasoning.[54] (3) Explanations must be phrased as general propositions to be modified or rejected if they fail to pass appropriate tests.[55] (4) Tests consist of examining predicted differences between experimental and control, reference, or standard data. However, (5) neither in logic nor in science can one prove the truth of general propositions, and in science one cannot even disprove them. For (6) general propositions in science can be tested only in conjunction with auxiliary propositions about accuracy, relevance, and "other things being equal." Thus, (7) only the conjunction can be proved wrong, which enables one to protect indefinitely one's pet hypothesis.[56] Though logically defensible, such procedure is frowned on in the best circles.

These philosophical problems are inseparable from the empirical problems, as will become clear in later chapters.

1.6 Analogy

Recognizing similarities is 'more valuable and less easy' than recognizing differences.

Wigglesworth[57]

"But Gussie isn't a parrot." Bertie Wooster on refusing to accept the analogy between an inebriated parrot and a human subject "unaccustomed to alcohol."

P.G. Wodehouse (*Right Ho, Jeeves*)

We use animals such as guinea pigs, white mice, rabbits, frogs, and fruit flies to gain knowledge we can apply to other organisms, including man. What's true of one species (or one individual) will not be true in all respects of others; one can only hope that different species (or individuals) are sufficiently alike in ways that matter. In choosing an experimental animal we therefore risk its being unsuitable, or at least having the critics say so. William Harvey had to put up with contemporaries who disparaged and ridiculed "with childish levity the frogs, snakes, flies, and other lower animals" whose circulation he studied; nor did they "abstain from scurrilous language."[58] Harvey's detractors have their modern counterparts, as witness the following item in the *Columbia Missourian* for November 17, 1994. Appalled at the thought of $85,000 being spent on watching the mating habits of guppies, the managing editor finally understood what serious science consists of. He wrote as follows:

> It's the bizarre projects smart people think up to avoid having to get real jobs. I know another, even more serious, scientist who has spent his professional career listening to frogs mate. He doesn't even get to watch. Of course, he's internationally famous among other serious scientists.
>
> Then there's the scientist, taller but no less serious, who drags himself down to Puerto Rico every winter to study songbirds. I'm not sure he cares if they mate. He's famous, too. Serious scientists appreciate each other. The rest of us can only watch in awe. . . .

The writer evidently thought these projects as valueless as trying to extract sunbeams out of cucumbers. I can sympathize with his difficulty, for it is indeed hard for a nonscientist to understand that science progresses by choosing rather simple-minded problems through which to understand broader issues. Perhaps, like Harvey's discovery of the circulation of the blood, some fundamental understanding of human behavior will emerge from the study of guppies, frogs, and songbirds. Only the future knows for sure, and for the present, such studies fulfill their function by teaching us more about the mysterious world around us. Trying to understand what scientists are up to requires more brains than poking fun at them.

After preliminary studies[59] on the health and parasites of the wood mouse, Charles began working on the short-tailed field mouse or vole (Figure. 1.7). Back in the 1920s it was by no means obvious that voles might be sufficiently like lemmings to be used to solve problems common to both, let alone those common to other populations, cyclic and noncyclic. It was known that in some places and in some years voles increased to plague proportions and just as mysteriously disappeared. It was not known whether they did so at regular intervals, or in synchrony, or by emigrating. But the analogy seemed close enough to be useful. And even if it were not, there were problems about voles that needed looking into. It was also good strategy to choose a species of economic importance, both in forestry and agriculture. At the start of this work almost everything had to be discovered, so it seemed best to start with the most puzzling of all components, namely, the decline or "crash" at the end of a cycle of abundance. Discovering differences between this and the other phases seemed the best way of understanding the whole cycle.

Differences between voles, lemmings, snowshoe hares, game birds, and noncyclic populations, especially in their social systems, may perhaps seem too great for there to be a common explanation for the behavior of their populations; but an explanation for one kind of population that applies to others will be preferred to a separate explanation for each. This is the principle of parsimony, often referred to as Occam's Razor, which states that one should choose the simplest of several hypotheses that, *at the time*, explains

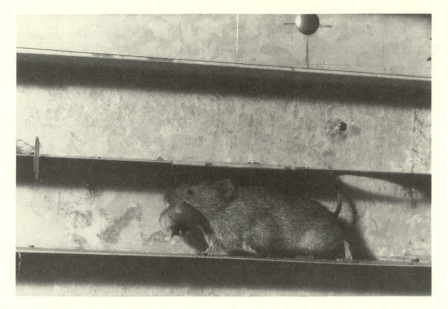

Figure 1.7. Short-tailed field vole, *Microtus agrestis,* carrying a nestling from one nest to another along a gallery, 2 in. × 2 in. cross section, similar to those in Figure 6.4. Being unable to count, the mother returns once more after removing the last nestling and continues to return as long as an observer keeps putting nestlings back where they came from. Photo by Denys Kempson.

the same observations equally well. This precept leaves room for argument[60] about the meaning and application of 'simplest' and 'equally well;' but bouncing ideas back and forth between populations is a good way of refining one's ideas or getting rid or them. Even though one's heart is pure, however, there's a risk of being accused of forcing one's ideas into a Procrustean bed, and one may need the strength of ten to accept the criticisms of one's peers. At first I camouflaged my views with the weasel words[61] that they had "implications which are at variance with some of the existing views about cycles in birds and mammals." The conflict remains, as will become apparent, and I have since been more explicit; for although I don't have the right answer, I don't know what's wrong with it.

1.7 No Selection, No Cycle?

At each mention of the Chitty Hypothesis faculty and students bow their heads and cross themselves to the accompaniment of religious music.
Stage directions for a student skit by Maura MacInnis and Jordan
Rosenfeld at UBC in December 1993

In this section I give an advance look at the ideas described in the rest of the book. They have become known as the 'Chitty Hypothesis,' which is unfortunate, because it's uncertain which stage of an evolving explanation should be so labeled and because Voipio[62] proposed a genetic explanation long before I did. I would prefer the hypothesis to be considered as originally stated,[63] namely, that "all species are capable of limiting their own population density without either destroying the food resources to which they are adapted, or depending upon enemies or climatic accidents to prevent them from doing so." Stating an idea in the form of a universally testable proposition is a device for suggesting experiments to falsify it or restrict its scope: it is a challenge to discover the truth, not a claim to know it. Having a bright idea, however, is of little use without suggesting how it works; and as early guesses are likely to be wrong, they furnish one's critics with reasons for throwing out the conceptual baby (self-regulation) with the explanatory bath water (how animals do it).[64]

In its simplest form this hypothesis applies to any species that limits its breeding density by means of territorial behavior. It also applies to species with subtler forms of spacing behavior, in which, at least once in its life cycle, individuals avoid, threaten, inhibit, or otherwise influence the survival, reproduction, or movements of other individuals: for example, by means of odors. Such spacing behavior, however, would tend to keep numbers stable unless associated with qualitative changes whose effects persist for one, two, or more generations. I once thought that such changes were purely adverse maternal effects,[65] but later suggested that the kind of animal present during the increase phase was selected against as crowding became more intense ('the polymorphic behavior hypothesis').[66] I believed that selection favored docile animals at first and aggressive animals thereafter. If anything, the reverse is true;[67] but even this may be irrelevant. It's best to assume that we have yet to discover "an unknown kind of interaction [that] produces an unknown change in the average properties of the individuals, whose descendants become more susceptible to unknown and principally local forms of mortality."[68] Thus, the first assumption of the Chitty Hypothesis is the prevalence of spacing behavior, or hostility: terms I use throughout the book as a cloak for our ignorance. The second assumption is the persistence of their effects on the quality of later generations, which may explain why cyclic populations go on declining for 1 or more years after they have become scarce and remain scarce in spite of a favorable local environment. Although the role of selection is doubted by some authors, I still think it likely that individuals are maximizing their fitness throughout the cycle.[69]

Figure. 1.8 gives the gist of these ideas. The curve *abc* shows the potential for increase among populations living in productive habitats. Sooner or later, however, populations stop increasing, which means that they recruit fewer offspring, exclude surplus animals from the breeding population *(cd)*, or in some way bring numbers back to their original level *(a)*. But instead of being confined to the start of breeding, losses may be distributed throughout the year in irregular incidents, as indicated by the line *aed*. The pattern *cd* is most easily seen among birds such as chickadees fighting for breeding territories on the break-up of their winter flocks.[70] Sudden reductions in numbers also occur among voles, and though their behavior is not observed directly, it is

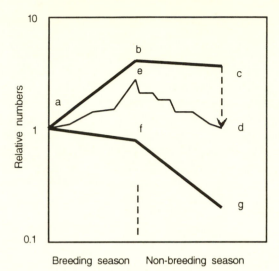

Figure 1.8. Generalized changes in numbers during transition from the maximum rate of increase (abc) before a cyclic peak to the maximum rate of decline (afg).

most parsimoniously attributed to a similar expulsion of surplus animals.[71] The curve *ae* is based on irregular recruitment of voles in the breeding season,[72] and *ed* on sudden losses of red grouse in the nonbreeding season.[73]

The curves *abc* and *afg* are based on patterns among snowshoe hares in Alberta.[74] During the decline from the first peak (Figure 1.9) numbers fell throughout the year until they hit bottom, as shown by *afg*. From 1966 to 1971 numbers increased between 1.5 and 2.4 times annually, after which they decreased faster in each of the 4 years after the peak, the proportions alive from one April to the next dropping from 0.48 to 0.17 at *afg*.

The first signs of regulation showed up right after the turn of the cycle, as 'realized natality' declined in each and every year from 1966 (17.8 young per female) to 1973 (only 7.5 young per female). I attribute this progressive decline to direct and delayed effects of spacing behavior on the length of the breeding season, juvenile survival, and litter size, all of which began decreasing while numbers were less than half what they were at peak density.[75] Winter food shortage, which is blamed for later reductions in reproductive rates,[76] can hardly be blamed at this early stage. The fate of fluctuating populations is probably sealed at the first signs of a reduced rate of reproduction, which may sometimes occur much earlier in the increase phase than I used to think. High numbers later on may merely modify the rate of a decline that's going to take place anyway.

My interpretation of the data on snowshoe hares differs from that of the authors, and the rest of this book deals with the struggle of my own hypothesis for survival against its well-established rival, namely, the doctrine of density dependence. Admittedly, there is abundant *a posteriori* evidence for density dependence, that is, that the severity of various mortality factors is correlated with population density. These are effects, however, and it's an assumption

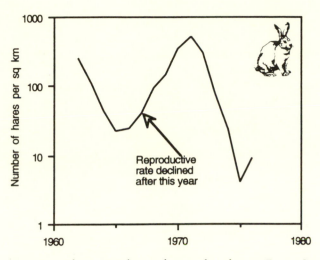

Figure 1.9. '10-year' cycle in numbers of snowshoe hares. From Cary and Keith (1979: Table 8).

that they are also causes of population regulation. This assumption has produced one of those systems which, "in its prime . . . is a triumphant success: in its decay . . . an obstructive nuisance."[77]—or at least I think so. However, a newly graduated student who mentioned that there was little evidence for density-dependent regulation caused one well-known ecologist to hit the roof. "Such appalling ignorance," he wrote, "is not dissimilar to believing that the world is flat, or that evolution is a myth"[78] Whereupon the writer fell with a resounding crash into the trap he was warning against, namely, the Infallibility Syndrome, an insidious affliction of middle-aged and retired scientists. "The trouble with theorists," says Crick, is that they "almost always become too fond of their own ideas, often simply by living with them for so long. It is difficult to believe that one's cherished theory, which really works rather nicely in some respects, may be completely false."[79]

Debates about density dependence generated more heat than light in the 1950s, and are still going on. For a recent appraisal of this doctrine the reader should consult Sinclair,[80] according to whom:

> Nicholson's theories of equilibrium [1933 and later] led in different ways to the works of both Wynne-Edwards (1962) and Lack (1966) and to the outburst of field studies. It is perhaps not appreciated that by 1957 Nicholson had anticipated most of the modern concepts on regulation; local extinctions, curvilinear density dependence and density-vague phenomena.

It is apparent from this review, however, that these modern concepts have no experimental justification.

Using the same data to support different kinds of tunnel vision is common among scientists, who in this respect are much the same as other human beings. Picking on some aspects and ignoring others guarantees a succession of stimulating debates. These are often divisive, but trying to evade the criticisms of one's peers (however trivial and misguided) is to violate one of the most important characteristics of the scientific process. Putting one's views on record is the scientific equivalent of offering up a sacrificial lamb, which is the purpose of this book. Population ecologists seldom pull their punches, and I expect no kinder treatment than any meted out in the following pages.

Pioneering observations, 1929-1939

I insist on the radically untidy, ill-adjusted character of the fields of actual experience from which science starts. To grasp this fundamental truth is the first step in wisdom, when constructing a philosophy of science.

Whitehead[1]

2.1 Collecting Field Data

Douglas Middleton, as the original middle dog in the Bagley Wood study,[2] became permanently saddled with the nickname 'Dog' and protested in vain against the omission of the 'u.' His main contribution to the vole work was in organizing the country-wide collection of data. As there was almost no money for anything as untraditional at Oxford as doing zoology in the field, Doug toured the country in a small car, living off a large ham and sleeping by the roadside. After several weeks of travel he had persuaded 38 volunteers, mostly workers on forestry plantations, to set lines of traps and send the catches to Oxford—a procedure to which His Majesty's Post Office took exception when a tin of putrescent voles got squashed in transit. He also studied the distribution of grey squirrels, worked with others on vole diseases, and collected records of fluctuations in game animals, especially partridges. On my first field trip with him, we were returning from a partridge shoot along a rutted track in a van crowded with beaters, keepers, and dogs. The men, in merry mood after a good lunch, pretended the van was a ship ploughing a rough sea. This was too much for Doug, who had looked on the wine when it was red. Professional etiquette forbade him to be sick over the new colleague on his left, and kindness to animals prevented his ruining the glossy coat of the black retriever at his feet. He therefore took the remaining option of bringing up his lunch over the britches of the head gamekeeper on his right. Doug had a genius for forming good relations with country people, whose help was vital to the early work of the Bureau,[3] but this was not his finest hour.

At a famous meeting of the British Ecological Society, Doug stated that his best correspondents were country gentlemen with a housekeeper and a couple of acres. Except for cognoscenti of British slang, early twentieth century, readers will have to guess why this remark had the audience rolling in the aisles. Charles himself now and then regaled our teatime visitors with this tale, but being innocent of its coarser interpretation, quoted the man's estate sometimes as five, sometimes as ten acres.

In 1929-1930 voles destroyed a million young conifers on a single Scottish plantation of the Forestry Commission.[4] These plantations had been started to ensure a home-grown supply of timber, which had been in short supply in World War I. They were among the few vole habitats sufficiently undisturbed to be useful for long-term research; and thanks in part to economic losses due to voles, the Forestry Commission provided modest support for a program of pure research. Collecting basic knowledge would have to precede any practical applications, and would not guarantee success in replacing the current ineffective methods of trapping and poisoning. One possibility (remote, as it turned out) was that forecasting peak years might enable a forester to postpone a planting program; but the war put an end to such schemes.

In 1929 there were no substitutes for anecdotal evidence about vole populations, so ways of monitoring them had to be invented. One way was by trapping: on most areas in September only, on others in April as well. Trapping ended on most plantations in 1935, on others in 1937, and on all of them in 1939 with the outbreak of war. Some of the areas are shown in Figure 2.1.

On each occasion 50 unbaited traps were set five paces apart in a single line, visited and reset on 5 successive days. The total catch provided an index of relative abundance (the "trap index"). Of 38 areas sampled at one time or another, 8 received fuller treatment; besides being trapped twice a year by the forester, they were visited twice a year by Charles or David Davis[5] and later by Helen and me. The drill was as follows. Holding aloft a strange device, consisting of a large circle of wire, the worker marched across country along a line of 100 to 175 stakes spaced 20 yd apart.[6] He or she then placed the circle round each stake in turn, and on hands and knees scratched around inside the enclosed 2 sq yd until the first fresh droppings were found or the whole area had been searched in vain. The resulting 'trace index' of abundance was based on the percentage of circles with fresh droppings. The advantages of this system were that over a distance of a mile or more the index covered good, poor, and patchy habitats in a wide range of plant communities, could be obtained in 1 day, and provided a check on the trap index. The two indices usually changed in parallel.

Three of the eight areas provided samples for reproductive studies throughout 1930 and 1931, two were used for studying the effect of voles on vegetation, and three for epidemiological studies. These first 10 years constituted the all-important natural history stage that must precede an explanatory stage, though there's seldom a sharp line between them. Ecologists who skip this lengthy stage do so at their scientific peril; they must be brave enough to suffer the slings and arrows of outrageous comments from their lab-bound colleagues that they are a bunch of old-fashioned naturalists.

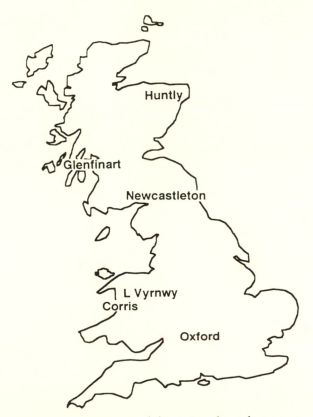

Figure 2.1. Distribution of the main vole study areas.

Unlike some studies, the vole work, before getting down to details, sampled a number of populations from a wide geographical range. It may seem that this is so elementary a precaution that no one would confine his attention to one area and so risk describing merely local events. The aim of every scientific study is to find the general in the particular; a wide survey is thus a form of insurance against the danger of mistaking the particular for the general. Unfortunately, there's seldom time or money for doing the right thing in ecology, and it may be years before the broad picture emerges—most likely as a collage of work by other people using other methods on other species in other parts of the world. They that have not patience should avoid field work. But they should be less patient than Charles, who never got around to publishing the full results, which appear for the first time in the next section.

2.2 Fluctuations

I must begin with a good body of facts, and not from principle, in
which I always suspect some fallacy.

Darwin[7]

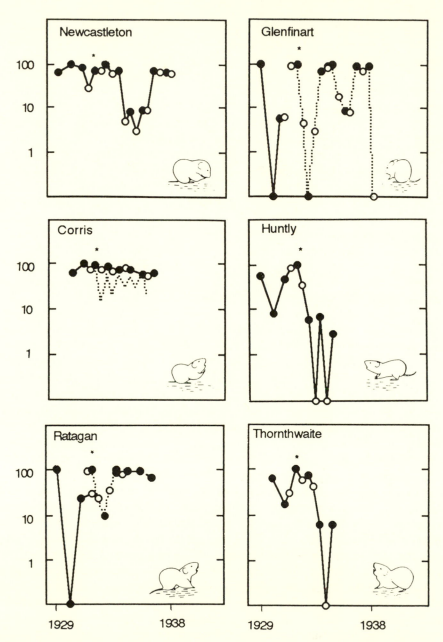

Figure 2.2. Relative abundance of voles in spring (○) and autumn (●) 1929–1939 according to the trap index (—) or trace index (....). Owing to changes in vegetation, some lines were abandoned or relocated (e.g. in spring 1932 and 1935 at Glenfinart, where peak values are not comparable). * shows 1932.

Data in this section were the first to be collected, but with two exceptions[8] remained in limbo until Helen and I rehabilitated them in the late1970s. Results from the trap or trace indices are summarized in two groups: in Figure 2.2 for areas visited twice a year (as a rule) for up to 10 years, and in Figure 2.4 for areas trapped in autumn only from 1929 or 1930 until 1934 or 1935. In most places, 1932 was a peak year or the final year of increase, but apart from this degree of synchrony (which could have been fortuitous or owing to plantations being of similar age), few populations fluctuated in step. That cyclic populations were outnumbered by noncyclic populations is of no significance, the frequency of the two types being contingent on Doug Middleton's choice of sites.

Populations in the first group had pronounced fluctuations in numbers except at Corris, where, according to the trap index for 1930-1938, the population fluctuated between spring and autumn only and so must be regarded as noncyclic. But according to the trace index, which ran only from 1932 to 1937, numbers were appreciably higher in September 1932 than in the next year. The slow decline from 1932 onwards was probably due to changes in habitat, especially in the growth of conifers.

I have particularly happy memories of one of these areas, Glenfinart, because its voles were out of phase with those at Newcastleton, where I had been introduced to the trace index during scarcity in autumn 1935. On this, my first field trip with Charles, I experienced the tedium of scratching around all day for nonexistent droppings and was also badly bitten by harvest mites. Population ecology had lost its glamor. Luckily, there were more than enough voles at Glenfinart, the next place on this historic journey, to make the work easy and exciting. In fact, as we approached the area Charles jammed on the brakes, and claiming he could find fresh droppings in 10 seconds flat, leapt from his car to the roadside, and won his bet with time to spare. I decided to stick with population ecology.

Voles at Glenfinart had reached peak numbers in 1922 and 1926 according to earlier information, and in 1929, 1932, 1935, and 1937-1938 according to Figure. 2.2. Scarcity was followed by immediate recovery, which in 1933-1934 and 1934-1935 was partly due to breeding in winter or early spring.[9] Voles on the nearby areas of Glenbranter and Ben Lagan were in phase with those at Glenfinart. The violence of these fluctuations was shown in 1929-1930, when the 5-day catch fell from 137 to 0. At Newcastleton, by contrast, the ups and downs were less abrupt than at Glenfinart. For the 4 years 1929-1932, voles remained fairly abundant each autumn, with only slightly lower numbers before and after a low peak in 1930. A higher peak in 1933 was followed by 3 years of decline to scarcity—through my arrival on the scene in September 1935 to a low point in spring 1936. Numbers peaked again in 1937 and had declined only slightly by spring 1939.

From data obtained 6 months apart an investigator will be unable to distinguish between uniform and uneven rates of survival and will have unrestricted freedom to speculate about intervening events. It became obvious

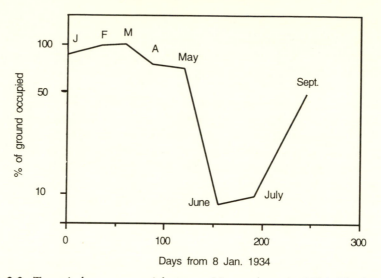

Figure 2.3. Trace index on a special area at Newcastleton in 1934. The September value is applied from the main area, where it was 71% of the April value. From Elton et al. (1934, Table 1).

early on that intervals of 6 months without intervening data, were useless except for showing peaks and troughs. As can be seen from Figure 2.3, additional monitoring showed that the rate of decline at Newcastleton was far from uniform: the trace index remained fairly high for at least 1 month after the regular visit in April, fell abruptly during the next 35 days, remained low for another month or so, and by September 1934 had returned to 71% of its fairly high April value. The trap index was even higher (Figure 2.2). If data had been obtained only twice this year, the most significant events of this decline would have been concealed.

Having dismissed the autumn recovery as "only slight,"[10] Charles correctly predicted the scarcity of 1935; but given the apparently strong recovery in September 1934, he did so for the wrong reason. More than a decade later, for what appeared to be the right reason, I wrongly predicted that a strong autumn recovery would be followed by abundance the next spring. And later still, I wrongly predicted that a similar recovery would be followed by scarcity. Such were the perils and dangers of those days; but they were helpful in showing that predictions that don't come off may be more valuable than those that do.

Five other areas were monitored in a similar way but for a shorter time; data for the first three are shown in Figure 2.2, data for the fourth, less well documented area, are shown in Figure 2.4, and data for the fifth are not worth plotting.

Huntly—Voles were at a peak in 1929 and again in 1932, when, contrary to expectation, numbers had gone up during the previous winter. After a high peak that spring, numbers remained low for the next 3 years.

Ratagan—There seem to have been peaks in 1929, 1932, and 1934 or later. Voles evidently became much scarcer in 1930 than in 1933. The trapping data for 1933 are missing, but the trace index was high enough in 1932 and 1934 to suggest a quick recovery from only moderate scarcity in 1933.

Thornthwaite—Voles were abundant in autumn 1930 and again in 1932, though they did not decline until after a second autumn of high numbers. They declined in summer 1934 and remained scarce for the next 12 months.

Rendlesham (Area 44 in Figure 2.4)—Voles were trapped twice a year for 3 years only. Numbers were highest in autumn 1933, about half as high in autumn 1934, and very low in spring 1935.

Lauder—Voles remained fairly abundant from 1929 through 1933, with slight peaks in 1929 and 1932. None were caught in 1934 and only a few in 1935. Apart from providing this contrast, the data are unreliable because of rapid changes in vegetation and the need to keep changing the lines on which the indices were obtained. I have not plotted these data, but refer to this area in Section 5.3.

Besides getting indices of abundance, Charles and David Davis estimated the density of the Newcastleton population by trapping out all voles from an area of 3000 sq yd (using 324 snap-traps) and concluded that in March 1934 there were 100-200 voles per acre.[11] If numbers went down at the same rate as the trace index (Figure 2.3), over 30% of the voles present in March were lost in the 60 days before the April visit. But between May and June about 90% of the voles were lost in 35 days, an acceleration of roughly five times. The loss from the whole area was between one and two hundred thousand voles. These estimates, crude and untrustworthy by present standards, were the first in a long series obtained by increasingly accurate methods.

Populations in the second group covered the range of patterns shown in Figure 2.4, which presents a sample from 30 sets of data obtained by local helpers from areas distributed across the length and breadth of Great Britain. Patterns varied from that for area 19, which had obvious peaks in 1929 and 1932 with 2 low years in between, to that for area 23, where catches remained fairly high for all 7 years. Catches on the rest of the areas fluctuated between these two extremes or declined slowly rather than suddenly and without displaying any common pattern. Without data for spring, however, we cannot be sure that these populations were noncyclic.

Thus, the rise and fall of some vole populations was just as dramatic as among lemmings, and at places had a similar periodicity. But at others the fluctuations were irregular or scarcely detectable. So it was easy to discard the idea that all vole populations had regular cycles in phase with one another.[12] Other ideas took longer to get rid of.

Most of these areas had been abandoned before I arrived in Oxford, partly because there was no time to monitor them closely, sometimes because the habitats were poor and numbers too low to give reliable results, but chiefly because

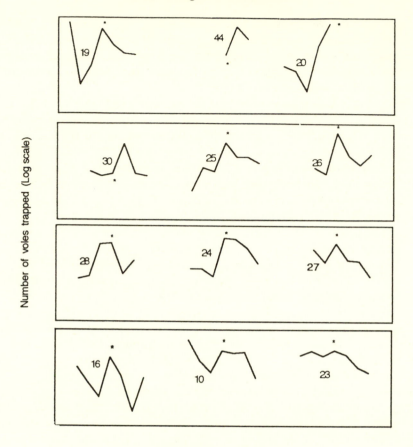

Figure 2.4. Number of voles trapped on miscellaneous areas in autumn 1929–1935. Area 44 is Rendelsham (spring catches omitted). * shows 1932. Values on the y-axis go from 1 to 100 in each panel (log scale).

the main study was restricted to habitats in which cycles were likely to occur.[13] To have studied fluctuations of all kinds would have spread the work too thin; but even superficial data provide an economical guarantee against assuming that what is true for a few selected areas is true for all others.

Natural phenomena such as cycles will occur if and only if certain initial conditions are satisfied. We do not know why they were satisfied at some places and not at others; but because (for example) Corris and Newcastleton are 250 miles apart and differed in so many ways, we could not tell which differences were relevant to differences in the behavior of the populations and which were not. And for other areas the numerical differences between cyclic and other types of decline were too slight and the biological data too poorly documented to be analyzed. Data limited to the autumn provide a different pattern from data limited to the spring, and neither tells enough of the story, whether alone or together.

Here, then, was a mass of natural history data that had to be explained, and the rest of this book shows how, for the next half century and more, one hypothesis after another came to grief.

Charles was writing up the results when war broke out, and only part of that material appears in *Voles, Mice and Lemmings*. His typescript includes details of the topography of the trap and trace lines, the vegetation, other species caught in the traps, and various factors affecting the interpretation of the results. Anyone wishing to get more out of the figures may consult the archives in the Elton Library, which as part of the Department of Zoology, Oxford University, provides another lasting tribute to Charles's contributions to ecology.

2.3 Reproduction

It seems that no attempt to solve a scientific problem can even be begun without the subsidy of some hypothesis, however dimly formulated or however vague.

Medawar[14]

Richard Ranson established and maintained a breeding stock of voles. He and John Baker, a member of the Zoology Department best known as a cytologist, made the first study of body weights and reproduction in voles. The flip side of their work was developing a chemical contraceptive,[15] which at first they tested on guinea pig semen; but according to a scandalous rumor, they had a human donor working for them in a locked room in the basement of a nearby building, Professor Goodrich having vetoed the use of his department for this research.

Lemmings and voles have the capacity to increase rapidly, so this phase, however spectacular, seemed understandable, and only vague hypotheses could be entertained about the relevance of reproduction to the cycle problem. Definite hypotheses could, however, be entertained about the relevance to reproduction of certain physical variables. To study the effect of locality, John Baker and Richard Ranson chose three areas that sampled a climatic gradient of 360 miles from north to south: from Huntly, through Newcastelton, to Corris. They looked for correlations with rainfall, temperature, and light, and provided data on body weight and reproductive condition[16] for 2500 voles for 2 years from January 1930 through December 1931 (Figure 2.5). They concluded that voles are seasonal breeders, apparently needing 100 hours of sunshine per month, which meant that breeding lasted from February or March to September or October. But there were anomalies.

To their surprise, they found that breeding started latest in Corris, the

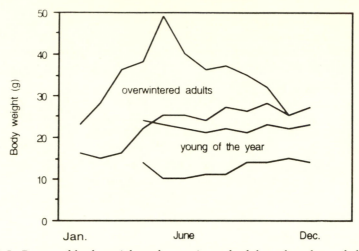

Figure 2.5. Range of body weights of overwintered adult male voles and their male offspring from three areas in 1930. From Baker and Ranson (1933: Figure 2).

most southerly of the three areas, and continued latest in the most northerly, at Huntly, where a few females were still fecund as late as December 1931. The other main conclusion was that young animals remained immature during their season of birth and overwintered in something analogous to a larval condition. This inference was based on a distribution of body weight, which became bimodal as breeding progressed, with a lower range around 20-25 g among adults, and a gap between them and the young, which had low body weights throughout the winter. These results helped determine the times at which to obtain the trap and trace indices. Most populations should have reached their seasonal maximum in September and decreased to a minimum about April. Like other short-lived animals, voles will decline in numbers during the nonbreeding season unless there's immigration, and declines of this kind must be distinguished from other kinds.

At the end of the 2 years, samples of voles continued to arrive at the Bureau, but twice a year only, and during 1932-1935 David Davis and Richard Ranson weighed each animal and recorded its reproductive state. But they had other jobs to do than analyze the data, nor was I asked to do anything about them. So the records lay around until the outbreak of war, when they were stored for safety in the basement of the Bodleian Library. Then, at odd times during the first 30 years after the war, Helen and I dusted them off to see what further information we could squeeze out of them. Our findings come later.

Charles soon discovered that a cyclic decline could continue in spite of breeding, and described as follows the course of events at Newcastleton in 1934: [17]

January and February, females not breeding yet, healthy; March, many females ready to breed (vulva perforate), mostly healthy; April, many females pregnant, epidemic [of disease] beginning; May, many females with young in nest, epidemic running its course, many young dying of starvation and probably also from the disease; June, most of the old and young gone, epidemic abated; July, August, September, slight recovery, the remnant breeding.

This quotation is from a paper written under the influence of the epidemic hypothesis and is a mixture of observation and interpretation. We can now dismiss the interpretation, but can agree that the decline was not due to lack of breeding.

2.4 Food

The most notorious source of fallacy is probably *post hoc, ergo propter hoc,* that is, to attribute a causal relationship between what has been done [or observed] and what follows . . . especially . . . in the absence of controls.

Beveridge[18]

Victor Summerhayes, a botanist at the Herbarium, Kew Gardens, spent his July holidays studying the impact of voles on vegetation. In 1932 he began a study of vole damage and continued until 1938, a period that included two peaks at Newcastleton and minor fluctuations at Corris. His method was to compare the state of the vegetation in protected and unprotected plots. Sheail provides a photograph of Summerhayes at one of his exclosures.[19]

An obvious explanation for a decline is that voles run out of food. As they often damage their vegetation, this explanation is reasonable and of botanical interest. Vole damage takes this form:

The voles cut up stems and leaves into pieces which are left lying either in the open or more generally in the runs, and which apparently are never eaten. In comparison with the amount eaten the amount left must be relatively small, but it nevertheless represents a considerable loss to the plants and gives a clue to the immense damage done . . . frequently the whole of an inflorescence several feet high may be found cut up into small pieces. . . . the voles gnaw away the tussocks of *Molinia* and other tufted grasses . . . Often the tussocks are found almost completely undercut.[20]

Voles also cut grass for nests, keep their runways clear by mowing down all sprouting vegetation, and bite the tips off grasses outside their runways. In July 1938 as many as 43% of the blades were without tips, though they were only a few centimetres shorter than those in the protected plots. The effect of vole damage was to "keep the [plant] community more open and thus to create a habitat favourable to the continued existence of a greater number of species." The dominant species nevertheless continued to thrive and maintain their frequency in the unprotected plots. Even during peak years the grass remained so thick that in places it "almost hid a small bota-nist." (The phrase is from Charles's words at a symposium;[21] he cleaned up his prose for publication, no doubt to spare the feelings of his diminutive col-league.)

At Newcastleton in 1934, voles remained abundant through May 7th (Fig-ure 2.3), by which time the new season's grasses were flourishing. If the sub-sequent decline was due to food shortage, the dominant vegetation should have been devastated by the time Summerhayes measured it in July. Figure 2.6 shows that it was not: between 1933 and 1934, two of the dominants had become more abundant and the third had remained about the same. Trends were similar in protected and unprotected plots.

Until 1938 the voles exerted no definite influence on the quantity of domi-nants in either forest, but in that year voles were at a peak at Newcastleton and did a great deal of damage. They did not, however, decline in numbers that winter, and were still abundant when the study ended in April 1939. Summerhayes believed the vegetation had become increasingly vulnerable to vole attack from 1936 onwards, all three dominants having declined during the last years of the study, inside as well as outside the protected plots. He attributed this to competition from roots of the conifers, which had grown considerably since 1932. At Corris neither voles nor vegetation changed in any obvious relation to one another. Summerhayes concluded that although the "effects of the voles on the vegetation are easily demonstrable . . . it has not always been easy to correlate the degree of vole effect with the fluctua-tions in the vole abundance. . . ." This conclusion came as no surprise to those of us who had groveled for hours on hands and knees in tall grass

Figure 2.6. Relative abundance of dominant grasses in unprotected plots at Newcas-tleton (similar in protected plots). From Summerhayes (1941). ↓ = decline of vole population (see Figures 2.2, 2.3).

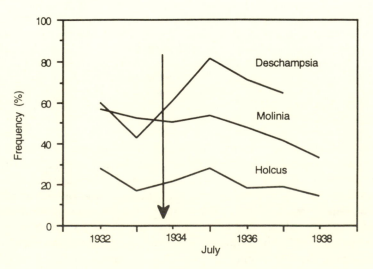

during times of high and low numbers of voles. Not only did populations decline when there was still plenty of food, but they sometimes failed to increase again until a year or 2 after all signs of damage had disappeared. Runways never occupied much of the area, the frequency of points without vegetation, even in peak years, being much the same outside as inside the protected plots.

As the dominants were only slightly affected, and other plants increased in diversity, voles were evidently not starving to death at Newcastleton or Corris, nor were they lacking a varied diet. Starvation, however, remained one of the ways in which declines might be modified, especially if drought and high numbers were to hit the vegetation at the same time.

Starvation was thus the first to be crossed off our list of possible explanations for cyclic declines in voles (and by analogy in cyclic populations of other species). We must therefore try to understand its long-lasting appeal to other workers. Some let their theories blind them to the force of contrary evidence, do not obey Darwin's Golden Rule or share his suspicion of principle, but prefer to trust to the belief that animals compete for limited supplies of food. For example, less food may indeed be available at the end of winter than at other times but may still be more than enough to satisfy the animals' needs.

Other observers are impressed by the association between damage to the herbage and subsequent decline of a vole plague. To someone who had driven (say) from Oxford to Scotland on purpose to see such a plague, the damage would be simple and convincing evidence of famine. As a random sample of events, however, such examples leave much to be desired; they are usually observed because of something exceptional. Correlations, however striking, are no substitute for experimental evidence. Moreover, the problem is not whether voles are capable of exhausting their food supply (they are), but why they seldom do so.

One more reason why starvation is a popular explanation for vole declines is that differences in amount or kind of food often account for differences in abundance *between* habitats—also within habitats when food is added or subtracted. Thus, numbers of lynx go up and down with the abundance of snowshoe hares; so do the numbers of birds and mice with changes in seed crops. *Changes* in numbers with *changes* in food supply, do not, however, imply that numbers increase up to its limits, and it is naive to assume that carrying capacity for wild animals is analogous to carrying capacity on a farm with limited resources. The natural spacing behavior of farm animals has been selected against under domestication. Furthermore, carrying capacity in nature is almost impossible to measure; it is not equivalent to the observed maximum population density.

Studies on small mammals are not the only ones in which the *effect* of high numbers on food supply is assumed to be the *cause* of a decline. The spruce budworm can destroy a high proportion of the foliage of its host tree, and "It is quite natural, therefore, that we frequently see the statement that outbreaks come to an end only because the budworm 'starves itself out.' "[22] Morris, however, was careful to distinguish between foliage depletion and food shortage or starvation and concluded that larval starvation was one of the least important causes of the collapse of an outbreak in the 1950s. In one of Morris's plots where defoliation was hard to detect, the budworm population became just as scarce as it did in a plot where "defoliation was very apparent and closely approached 100% in four of the epidemic years." The rate of decline, as might be expected, was greater where the decline was modified by damage to the host tree (Figure 2.7). Color photographs of these two plots show that one was green and one was grey. Anyone who had had the misfortune to study only the latter would have been tempted to commit the fallacy of *post hoc, ergo propter hoc*.

Committing this fallacy is a mistake that has bedeviled population ecology since its inception. Accurate field observations are, of course, an essential starting point, and looking for correlations is the next step. But because it's easy to invent explanations after the event, none can be trusted unless it's survived a rigorous search for contrary evidence. The next section includes a preliminary look at one way of finding it.

A useful resort to modeling is that of Fischlin and Baltensweiler,[23] who show in Figure 2.8 that there's enough agreement between their model and the real system to encourage their belief that the budmoths had ruined the quality of their food. There's also enough disagreement to encourage my belief that their model should be rejected for failing to account for a characteristic feature of real cycles, namely, the tendency for declines to continue for no apparent reason after the animals have become scarce. In the present case

Figure 2.7. Number of spruce budworm larvae per 10 sq ft of balsam fir. Both populations declined to extreme scarcity in spite of differences in food supply. Defoliation was almost complete in one stand, less than 10% in the other. From Morris (1963: Plate 1.1 and Figure 33.1).

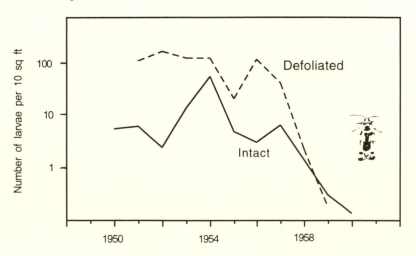

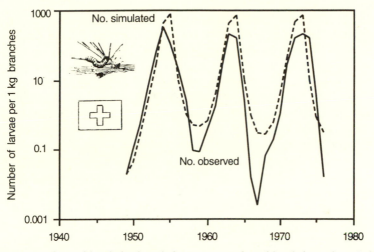

Figure 2.8. Number of larch budmoth larvae per 1 kg of larch branches. Solid lines show the mean number observed, broken lines show the number modeled. From Fischlin and Baltensweiler (1979).

numbers continued falling even when densities were less than 0.01% of peak density and poor quality of food was no longer a plausible option. The value of the model, in my opinion, lies not in the resemblance between it and observation, but in the discrepancy between them, which makes sense of the collapse of the cycle in 1990 without the predicted defoliation.[24] Dr. Baltensweiler, however, blames unusual weather in 1989-1991 for this departure from prediction and sticks to his view that "larch bud moths are unable to limit their own numbers except through the effects of damage to their food supply."[25] Nor had he been otherwise convinced by failure to prevent a decline on an area protected from defoliation. Given the "evidence that a population may collapse below a density level inducing food shortage,"[26] I deny that declines can be blamed on poor qualtiy food.

Dr Baltensweiler and I are likely to defend our own beliefs until one of us is hit over the head by the experimental equivalent of the legendary 2 x 4.

2.5 Predators

In this way [by looking at a single example] they merely attempt a universal syllogism on the basis of a particular proposition (like those who think they can construct a science of politics after exploration of a single form of government, or have a knowledge of agriculture through investigation of the character of a single field).
William Harvey[27]

There is also another class of philosophers who, having bestowed much diligent and careful labor on a few experiments have thence made bold to educe and construct systems, wresting all other facts in a strange fashion to conformity therewith.

Bacon[28]

A vivid memory of prewar work with Charles is the discovery, in 1936 at Glenbranter, of a long-eared owl roost from which the birds, with one dead exception, had departed after the disappearance of their prey. The ground, thickly strewn with pellets, showed the route through which hundreds, perhaps thousands, of voles had been recycled. Judged superficially, this association between decline and predation is an example of cause and effect; and later observations would probably have confirmed such an opinion if they had been confined to this area—an example of pseudoreplication.[29] However, one of the strengths of the prewar work was that not all eggs were in one basket. At Newcastleton, for example, there were no plantations with roosts suitable for long-eared owls, and apart from short-eared owls and kestrels, few other predators, such as weasels, were known to be present in most years. Yet voles declined in numbers anyway, apparently from epidemic disease.[30]

We now had a choice of two hypotheses: either that the decline at Glenbranter was due to predation and that at Newcastleton to disease, or that both were due to the same factors. The first alternative was based on differences, the second on similarities and the faith that all cyclic declines are alike, however modified by local differences in predation, disease, weather, food, cover, and other factors. But we obviously needed more data on predators, and Glenbranter seemed a good place to get it. Accordingly, besides setting up new lines for trace and trap indices, we labeled a number of posts at which the local forester collected owl pellets each month. We also enlisted the help of the Rev. J.M. McWilliam, who drove over now and then from Glasgow to observe the hawks and owls. The site of this proposed long-term study (Figure 2.9) was "a hill called Ben Lagan, which forms a massive natural unit some five miles round at the base, and about 1,600 feet high. Here a whole cycle of voles was measured in the standard way, up to the peak in 1938 and crash in the following year.[31] Alas, this work was interrupted by the war and never resumed.

At Newcastleton we had better luck with an attempt to understand the role of predation. We knew that no more than four pairs of short-eared owls were nesting on an area of 2020 acres;[32] and from studies on a captive owl[33] I obtained a rough figure for the number of voles the wild owls were likely to remove. I concluded that in April 1934 their total requirements would be about 50 voles per day or only 0.02% to 0.05% of the vole population, which was then between 100 and 200 per acre.

I left the argument at that point but will now expand it. To be on the safe side, let us suppose that all species of predators, including broods of short-eared owls, were together feeding at over 20 times the rate above. Then, if there were 100 rather than 200 voles per acre and no other prey, the owls could have eaten all 1000 of those that disappeared between March and

Figure 2.9. Ben Lagan, Glenbranter, Scotland in 1939, showing young conifer plantations and one of the unplanted corridors in which trap- or trace-lines were run. Photo
by Charles Elton, courtesy of Joy Elton.

May. This may be credible; but we next have to explain why predation increased between May and July to roughly 3500 voles per day. Such an increase would not be credible without evidence of a sudden change in predation pressure—for example, from an invasion of hawks, owls, foxes, etc. No
such invasion was apparent.

Thus, if the declines at Glenbranter and Newcastleton belonged to the
same class of events, one could rule out predation as irrelevant to the occurrence of cyclic declines; it would instead be one of many modifying factors.
But there was much to be said for the alternative view that sometimes predation, sometimes disease is sufficient to cause a decline. To rule out this possibility directly (as opposed to the doubtful calculations just offered) we needed
control areas where voles remained abundant in spite of heavy predation. We
had no such data, and what follows is a preliminary look at two related
problems discussed in Sections 2.9 and 5.8 in more detail than some readers
may wish to bother with. Instead, the first problem can be summed up as

follows, where C = a possible cause (predation by owls), E = the effect to be explained (a decline), and absence or scarcity is denoted by C′, E′.

	Many owls		*Few owls*
decline	CE	↖	C′E (Newcastleton)
no decline	CE′	＼	C′E′ (Glenbranter)

At Glenbranter we had an association between C and E, probably preceded by scarcity of owls in the previous year (C′E′). At Newcastleton we had a decline despite the scarcity of owls (C′E). To decide between conflicting conclusions about the relevance of owl predation we needed evidence of the missing type CE′: plenty of owls but no decline. A search for control evidence of this sort is a *sine qua non* in testing any hypothesis. To rely on mere correlations such as that at Glenbranter is to risk mistaking effects for causes and violates the most characteristic aspect of scientific procedure. Conclusions unsupported by controls are unacceptable.

The second problem at the time was lack of detailed evidence that cyclic declines are alike in one or more essential ways. Only if this is so are we justified in looking for one or more common antecedents. To replicate our results we must know when we are dealing with a specific, recognizable class of events and when we are not. For just as there is no one cause of every disease, there is no one cause of every decline in numbers. But as doctors must distinguish (say) typhoid from typhus, so must ecologists distinguish one class of decline from another. Only from knowing a patient's symptoms can a doctor diagnose the disease he has to treat; a mere high temperature is compatible with many conditions. And only from accurate natural history and experimental data can a scientist recognize the specific phenomenon he has to explain; a mere decline in numbers is compatible with many explanations. The tendency for loss of voles to recur at regular intervals of 3–4 years was the familiar feature of cycles; but better diagnostic criteria were needed, especially when there were no previous records.

Harvey rightly warned against jumping to conclusions "through investigation of the character of a single field." In spite of this warning, however, a single instance may sometimes be enough to put one on the right track. Indeed, "the actual procedure of science is to operate with conjectures, to jump to conclusions—often after one single observation."[34] Such conjectures are justifiable if, but only if, followed by a search for evidence against them. The difficulty for population ecologists is that it may be years before they can obtain such evidence. And since the original conjecture may have had a long and painful gestation, progress is bound to be slow and publication less frequent than in other branches of science. Meanwhile, half-baked ideas get into the literature and are hard to get out. "It is well known that demonstrating an error demands more time than committing it."[35]

The belief that predation explains the spring decline of vole populations

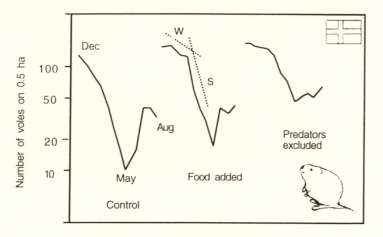

Dec. 1983 - Aug. 1984

Figure 2.10. Number of voles on three concurrent grids in southern Sweden. The change from good survival in winter (w) to poor survival in spring (s) is most marked on the grid with surplus food, but also occurs on the other two grids. Data from Erlinge et al. (1990, Fig. 4); icon courtesy of Jep Agrell.

has continued for the intervening half century, bolstered by faith in venerated theories about the powers of predators to reduce the numbers of their prey.[36] Figure 2.10 jumps from the 1930s to the 1980s, from Scotland to Sweden, and from poor to excellent data. Erlinge et al.[37] compared losses in the presence and absence of predators and showed, as might be expected, that voles disappeared faster when exposed to their enemies and reached a lower density. Unexplained, however, was how rates of loss accelerated in late winter, not only when predators were absent but also when extra food was present. Erlinge et al. provided no explanation for these increased rates of loss; indeed, they seem to have overlooked the magnitude of the change after February 1984, perhaps because they plotted their data arithmetically. In this form, as already explained, changes in rate are less apparent than when data are plotted in logarithms.

Their conclusion that "the dynamic pattern . . . seems to be determined by predation and probably food" seems inconsistent (to say the least) with their evidence that the populations declined in the absence of normal predation. The authors defend their hypothesis by saying that the decline on the fenced area "may have been due partly to weasel predation," which they reduced but were unable to prevent altogether. Nevertheless, this area lost slightly more animals than did the control. The evidently unpalatable truth was that the treatments failed to stop the declines.

My own colleagues, in a similar experiment on snowshoe hares, also failed to do more than delay a decline. They provide understandable evidence that densities increased 11-fold through an interaction between (permeable) fencing, feeding, and reducing predation. They have yet to explain why the decline proceeded in spite of abnormally high numbers and reduced predation. No doubt they will think up something. I have been markedly inept at brainwashing.[38]

To return to the Swedish work: voles disappeared faster, both on control and predator-exclusion plots, just before breeding started in 1984. Numbers fell even further in 1985 (see Figure 5.7), and the authors claim that "The change in numbers over these two years followed a pattern that has been observed in this area for many years."[39] If so, this supposedly noncyclic population provides a good example of a cyclic decline of Type H—low spring numbers in spite of recruitment the previous autumns.

While showing that the *dynamic* patttern, that is, spring declines, cannot be explained by predation or food shortage, these studies show that by making the habitat more favorable the authors were able to produce higher numbers and maintain them somewhat longer than on the control. If the habitat had been improved yet further (with better weather, additional cover, or different vegetation), the spring decline might have been modified even more, perhaps postponed for another year, as with the snowshoe hares. Improving a habitat alters the relation between a population and its environment, so that a formerly 'high' density now becomes 'low' in relation to the new conditions. Differences in abundance associated with differences *between* habitat types should not, however, be confused with the phenomenon of changes *within* a given habitat type.

Unlike my colleague Mick Southern,[40] whose primary interest was in predators, my students and I felt they were of marginal relevance to an understanding of cycles. This view received later support from a questionable discipline, namely, the wrong application of modeling with which population ecology continues to be plagued. In analyzing some earlier Swedish work, Kidd and Lewis[41] dispute the claim of Erlinge and his co-workers to have shown that predators can regulate their prey. Kidd and Lewis believe that "the regulative significance of vole predation can become apparent only with the use of simulation models incorporating realistic predation functions based on data from several years." In other words, when in doubt appeal to a model, not to old-fashioned criteria such as experiments and 'forcible' negative evidence. Moss and Watson[42] are right to point out that

> Too much reliance is being placed on unsubstantiated models, which are beginning to replace facts in the thought processes of too many ecologists. . . . Attractive models may seduce the scientist away from the mundane business of documenting long, slow processes.

2.6 Weather

It is the peculiar and perpetual error of the human intellect to be more moved and excited by affirmatives than by negatives; whereas it ought properly to hold itself indifferently disposed toward both alike. Indeed, in the establishment of any true axiom, the negative instance is the more forcible of the two.

Bacon[43]

The tendency for independent populations to be in phase suggests (to me at any rate) that weather is the most likely synchronizing agent. Early in his career Charles was impressed with the similarity in length of the sunspot and snowshoe hare cycles and suggested the one might be dependent on the other;[44] but he gave up the idea on finding that they cycled independently (an example of a forcible negative instance that did not discourage later authors from revamping the idea). He then began searching for local cycles in weather and was encouraged by the results of a study by A.H. Goldie, who "found that pressure-gradients, calculated in a certain way, showed a well-marked recurrent change since 1903 with a periodicity of three or four years."[45] Charles hoped that this discovery would speed up the search for other climatic cycles, but nothing came of it, and I was under the impression, perhaps wrongly, that he had quietly abandoned his belief in their relevance. For although populations often did march in step, they did so less closely than he had believed.

Changes in numbers associated with changes in weather are easy to find after the event and should be treated as mere coincidences until confirmed by later studies. So while the role of weather was and still is uncertain, any explanation for cycles has to be consistent not only with the tendency for them to be synchronous, but with their being so often out of step. According to conventional views, weather is a density-independent factor, that is, one whose effects are the same at all population densities. Killing the same proportion of two out-of-phase populations would therefore do nothing to change their relative abundance. So there were theoretical difficulties in seeing how weather could bring aberrant populations back into step with the rest. One could only hope that some new theory or lucky observations would come along. To a limited extent, and in their own good time, both hopes were fulfilled.

2.7 The Epidemic Hypothesis

When a species, owing to highly favourable circumstances, increases inordinately in numbers in a small tract, epidemics—at least, this seems generally to occur with our game animals—often ensue . . . But even some of these so-called epidemics appear to be due to parasitic worms, which have from some cause, possibly in part through facility of diffusion amongst the crowded animals, been disproportionately favoured.

Darwin[46]

In the 1920s there was every reason to believe that the cycle in numbers of lemmings and other rodents was due to recurrent epidemics of infectious disease. The first attempt to test the idea[47] was carried out in 1925–1928, mainly on wood mice in Bagley Wood, and seemed to confirm what Darwin had written (above). Besides becoming heavily parasitized, mice taken into captivity began dying off at the same time as those in the wild. Although no pathogenic organisms were recovered, symptoms of death resembled those expected from a neurotropic disease.

This pioneering study set the pattern for investigating the vole cycle. The plan was to monitor field populations, keep voles in captivity until they died from diseases brought in from the field, do post mortems, and culture the pathogenic organisms. In two such studies,[48] in 1933 and 1934, the voles apparently suffered an epidemic of *Toxoplasma,* a protozoan parasite attacking the brain. Half the voles died within a week of arrival, often in convulsions similar to those observed in the Bagley Wood study. These symptoms, together with increasing death rates among both captive and wild voles, were taken as evidence in favor of the epidemic hypothesis and encouraged Charles to raise money to build a small lab at Oxford, where work began in 1936. Concurrently, Professor Edward Hindle, at Glasgow University, carried out a study on voles shipped across the Clyde from Glenbranter. He found no evidence of *Toxoplasma* during a decline in 1938–1939, nor did it show up in the voles shipped to Oxford.[49]

We now know several things that were wrong with the methods first used to study the supposed epidemics among vole populations. Three of these were unavoidable: live-trapping techniques, including the need to supply succulent food, were still primitive; too little time could be spent monitoring the populations; and trapping and transport of the animals, as well as their care in London,[50] had to be delegated. When I began my own studies I had enough time for field work and seeing to the animals' trapping and transport to Oxford.[51] This meant that I could diversify the areas from which the animals were collected and take charge of them between capture and arrival at the lab. The result was a better set of data on which to base inferences about the relevance of disease to the decline in numbers. And once doubts began creeping in, the case for the epidemic hypothesis began to crumble. With some trepidation I explained why I thought the authors had selected the observations that supported their hypothesis and had neglected those that did not.[52] In hindsight the objections were as follows:

1. High losses in the traps (about 40%) could have been due to lack of succulent food in the nest box.
2. Increasing losses in storage and traveling could have been because boxfuls of strange voles of mixed sexes had become more aggressive in the breeding season than they had been in January or February
3. Unfavorable conditions prior to arrival in London might explain the high death rate in the laboratory.
4. The proportion of voles infected was slightly lower in samples from Glenfinart, where there had been a severe decline, than from Lake Vyrnwy, where there had not.

5. Uninfected animals died as fast as those with *Toxoplasma*.
6. According to other authorities, the organism was not *Toxoplasma* anyway.[53]

I've recently thought of another reason for the high death rate in captivity, one that I find more plausible than item 3. It goes as follows: in the 1933 study four shipments of voles were received in London at intervals of 8–10 days. Besides dying off fast, these voles died in groups. In a representative sample of 16, for example, the distribution of deaths was as follows:

Days after arrival	5	6	7	8	9	10	11	12	13	14	15	16
No. of deaths	2	0	4	2	7	0	0	0	0	0	0	1

Similarly clumped deaths also afflicted wild voles sent to Oxford: "Occasionally both wild and laboratory-bred stocks died suddenly and in some numbers, often with liquid or gaseous contents of the gut."[54] No pathogens were found. The main breeding stocks never suffered these mysterious losses; they were housed elsewhere under the care of Richard Ranson.

After the war, when I started keeping my own voles, I found the factor most hazardous to their health was a clogged spout in the water bottle. If deprived of water for 24 hours the animals died with symptoms like those "which strongly suggest cerebral irritation."[55] As well, newly captured animals needed apple or carrot to tide them over until they learned how to use a water bottle. Once aware of these dangers, I was able to keep voles alive for months, regardless of the stage of the cycle. I therefore suspect that the clumped deaths in the lab were due to the technicians having refilled several bottles on the same day without checking that water could run freely out of the spouts.[56] This suggests that deaths in captivity, including those in the original Bagley Wood study, were artifacts and did not, as supposed, reflect the progress of an epidemic. The mysterious deaths described in the previous paragraph were also, I now believe, due to lack of water.

A more serious objection to the authors' conclusions is that Koch's postulates were not satisfied. In 1884, Robert Koch explained the steps to be taken to "satisfy oneself that the parasite and the disease are not only correlated, but actually causally related."[57] The pathogen must (1) be isolated from the diseased animal and grown in pure culture, (2) be injected into a healthy animal, (3) reproduce the original symptoms, and (4) be recovered from the sick animal. Koch's postulates provide one solution to the classical problem of distinguishing between correlation and causation. Bernard's 1865 'counterproof' provides another. In Sections 2.9 and 3.1 I suggest how both can be adapted to contemporary problems in population ecology. Meanwhile, the tendency to mistake correlation for causation remains a fault that won't go away. Baltensweiler and Fischlin have this to say[58] about the authors of an article,[59] published almost 50 years after the prewar vole work, in which a mere correlation was parlayed into a causal relation:

In 1957 . . . it seemed obvious to everybody that the granulosis virus disease played a crucial role in suppressing the outbreak [of larch budmoths]. But since substantial numbers of diseased larvae could not be found in subsequent outbreaks, this idea had to be abandoned. It is unfortunate that this fact was not taken into account in recent theoretical papers that revive the pathogen–host relationship hypothesis for budmoth cycles. It is worth noting that if the larch budmoth project had been terminated after the first 10 years of research, we would not have discovered that disease epizootics were not a regular phenomenon, and most scientists would now be convinced that the cycles were caused by the interaction between virus and budmoth.

Modeling did no damage on this occasion, but the same cannot be said about other models,[60] though the trouble comes from readers who accept mathematical arguments as a substitute for scientific evidence instead of a suggestion to get it. The inference that "parasites are sufficient to generate population cycles [in red grouse] . . ."[61] cannot be justified merely from finding that birds suffering from strongylosis survive less well than those that do not.

The next section describes the final nail in the coffin of the epidemic hypothesis—except that nothing is ever final in science.

2.8 Vole Tuberculosis

I have steadily endeavoured to keep my mind free, so as to give up any hypothesis, however much beloved (and I cannot resist forming one on every subject), as soon as facts are shown to be opposed to it.

Darwin[62]

I had, also, during many years, followed a golden rule, namely that whenever a published fact, a new observation or thought came across me, which was opposed to my general results, to make a memorandum of it without fail and at once; for I had found by experience that such facts and thoughts were far more apt to escape from the memory than favourable ones.

Darwin[63]

Dr. A.Q. Wells, M.D., D.M., gave up a successful career in London for the pleasanter life of working in Oxford in his own small lab at the back of the Sir William Dunn School of Pathology. He was a kindly but majestic figure none of us called by his first name.

In 1936–1939 I carried out the first of a long line of mark-recapture studies[64] and, in association with Dr. Wells, continued to test Charles's epidemic hypothesis. My part of the job was to collect field data and supply voles for pathological studies. The results of this work were twofold: Dr. Wells discovered vole tuberculosis, and we all abandoned the epidemic hypothesis. Dr. Wells announced his discovery as follows:

> Griffith (1930) states that tuberculosis in warm-blooded animals living wholly in the wild state is unknown. The object of this paper is briefly to report a widespread occurrence of tuberculosis in voles. . . .[65]

Dr. Stanley Griffith, though miffed by the abrupt contradiction of his claim that tuberculosis was a disease of domestication, undertook to type this new organism and do further work on it. As vole tuberculosis had possible clinical uses, Dr. Wells directed his research into this new and unforeseen direction:

> The existence of tuberculosis in wild animals may have some importance in the spread of the disease to man and domestic animals; the use in the laboratory of an animal which naturally contracts the disease may be preferable to the use of the animals, used hitherto, which are not known to have the disease in nature; and the fact that a small animal, easily maintained and bred in captivity here, is available for epidemiological study may throw some fresh light on tuberculosis.[66]

Thanks to sampling from a number of independent populations, we were able to show that vole tuberculosis was only sometimes associated with a decline in numbers and at others was rampant when they did not decline.[67] The disease was therefore irrelevant to this aspect of the problem. Although helping to put an end to the epidemic hypothesis, the discovery of vole tuberculosis vindicated Charles's belief in the value of studying disease in wild animals. Apart from its contribution to pure knowledge, this discovery led to the clinical use of vole tuberculosis to immunize people against the human strain. Also, until this most plausible explanation had been exorcised no better one was likely to be invented. For although it sounds good scholarship to lavish attention equally on several hypotheses at once ('The Method of Multiple Working Hypotheses'),[68] the advice is impracticable except in the early, superficial stages of a study. Only by ringing out the old had we rung in the new, for the specifics of the decline made it unlikely that disease of any kind could, by itself, produce the characteristics of a decline in numbers and forced us to rethink the whole problem. This reappraisal was incomplete when war broke out, and Charles voiced his doubts as follows:

> Neither of the two diseases so far encountered throws any light on the *regular recurrence* of vole-crash periods . . . they are seen either as local and incidental consequences of high density, or else as phenomena not primarily linked with density . . . there is one disease (from *Toxoplasma*) which is reported in three different cycle crashes and is not seen again; and another (from an acid-fast bacillus) that . . . does not fit at all clearly into the cyclical picture. . . . We are [therefore] forced to look in other directions for an explanation.[69]

Looking in other directions kept us busy during the postwar years.

2.9 Inductive Elimination

> But what the scientist can often do with complete logical precision
> is to disprove hypotheses.
>
> Medawar[70]

> In point of fact no conclusive disproof of a theory can ever be
> produced; for it is always possible to say that the experimental
> results are not reliable, or that the discrepancies . . . are only ap-
> parent. . . . I have been constantly misinterpreted as upholding a
> criterion . . . based upon a doctrine of 'complete' or 'conclusive'
> falsifiability.
>
> Popper[71]

We now turn to the formal argument against the relevance of tuberculosis to
the vole cycle, assuming for the time being that cyclic declines belong to a
single class of events. In the following discussion such declines are the effect
to be explained, and vole tuberculosis is the supposed cause; but the same
principle of exclusion or inductive elimination applies to testing other
hypotheses. The present section extends the discussion of the problem introduced in Section 2.5.

Part of the evidence in favor of any hypothesis is that the phenomenon or
effect (E) is associated with or preceded by some supposed cause[72] or causal
conditions (C). As a rule it will also be known that in the absence of the effect
(E') the supposed cause is also absent (C'). We then have the conjunction CE
and the disjunction C'E', which can be entered in the following 2 × 2 contingency table. If + means present (or pronounced) and − means absent (or
less pronounced), the evidence in favor of the hypothesis is as follows:

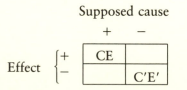

In terms of the epidemic hypothesis, C'E' might mean that voles were
healthy and survived well the first year, and CE that they were diseased and
declined in numbers the next; or better still, that in the same year, one population was healthy and one was unhealthy. Correlations of this kind are frequently observed, and explain why Charles Darwin supposed that epidemics
might explain declines in numbers, why Charles Elton set out to find out if
they did, and why, much later, Charles Krebs agreed they did not.

In general, the evidence that C causes E is unacceptable unless confirmed
by evidence that when C is removed or absent, E is also absent. "Feeling for
this necessary, experimental counterproof constitutes the scientific feeling *par*

excellence," says Bernard.[73] To avoid the fallacy of *post hoc, ergo propter hoc* we must rule rule out (a) mere coincidence and (b) the possibility that C and E are invariably associated through some common antecedent condition. We can often do so by examining enough instances of E to discover whether C is always associated with it. If we find the two are only sometimes associated, we say that C is not necessary for E, which is symbolized by the entry C'E in the upper right-hand corner of the diagram below. Such a discovery would save years of barking up the wrong conceptual tree. Dr. Wells and I soon found out that vole populations declined whether or not they had tuberculosis.

We nevertheless had to consider that although tuberculosis might not be necessary for a decline, it might be sufficient. This is the classical problem that an effect may have more than one cause. Disease may not be a necessary condition for a decline (C'E) but may be one of several factors that can produce it (CE). We can reject this possibility by finding a diseased population (C) that does not decline (E'), in which case C is not sufficient for E, and we enter the disjunction CE' in the lower left-hand corner of the diagram. As some tuberculous populations remained abundant, we concluded that tuberculosis was neither necessary (C'E) nor sufficient (CE') for a decline, and the evidence of its irrelevance now looked like this:

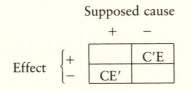

As inductive elimination plays so crucial a role in science, we must take account of Charles Darwin's *cri de coeur:* "My error has been a lesson to me never to trust in science to the principle of exclusion."[74] The trouble is that the method is designed for the elimination of error, not for the discovery of truth. Bacon, and more recently Platt,[75] believed that having got rid of false leads, only the truth would remain. Unfortunately, relevant factors will remain if and only if someone has already thought of them. "The success of induction depends on getting the right idea to start with."[76] Darwin did not have glaciation on his list of possible explanations for the "parallel roads of Glen Roy."[77] Having ruled out one of only two known possibilities, Darwin wrongly concluded that the "roads" were remnants of seashores, not of freshwater (i.e., glacial) lakes. "Physical discovery would be very easy work," says De Morgan,[78] "if the enquirer could lay down his this, his that, and his t'other, and say, 'Now, one of these it must be; let us proceed to try which.' Often has he done this, and failed; often has the truth turned out to be neither this, that, nor t'other."

Although guiding one's decisions, contrary evidence is never conclusive in science; Medawar (above) is wrong to claim otherwise.[79] Only in the abstract world of logic[80] does it follow that if a proposition p has implications q that are shown to be false (q'), then p also is false (p'):

$$\text{If p then q}$$
$$q'$$
$$\therefore p'$$

But in the real world of natural science propositions are never tested in isolation; they are always tested in conjunction with auxiliary assumptions about the accuracy and relevance of the observations. So now, if the consequences are false, then either p is false or something is wrong somewhere else. If k stands for these auxiliary assumptions, it's the conjunction of propositions (p.k) that is false: either p or k is false or both are false (p'.k'):

$$\text{If (p.k) then q}$$
$$q'$$
$$\therefore \text{(p.k)}'$$

If his own experiment has failed to come up to expectation, the author can protect his hypothesis by pointing the finger at k, the weather or other abnormal circumstances, unforeseen technical glitches, unexpected predation, breakdown of underlying assumptions, or other ways in which he can dismiss his own contrary evidence. Experiments in population ecology are so difficult to control that there are plenty of ways of defending one's hypothesis. And if someone else's experiment has confirmed a rival point of view, there are plenty of ways of attacking it. In addition to the *ad hoc* arguments above, the author may persuade himself of his rival's incompetence, stubbornness, ignorance of scientific principles, biased judgement, underrating facts that don't fit, selection of data, and failure to realize that anyone with ingenuity can explain an event after it has happened. Defense and attack are both justifiable, are indeed desirable, until an experiment has been replicated or the controversy has been settled in some other way.

The most common fault is for authors to cling to their pet ideas come what may (I myself may turn out to be one more victim); others, more rarely, give up too easily; they should at least make sure the contrary evidence is repeatable and should hang in there until something better comes along. Judgment plays a larger part in science than is popularly believed. (The symbol k is also known as the *ceteris paribus* or "other things being equal" clause. It's an escape clause that's sometimes legitimate, but only sometimes.)

Claude Bernard[81] and Whitehead[82] argued long ago that one's reasoning is the same whether evidence is induced or merely observed. For example, Pascal was experimenting when he "merely" observed the difference in barometric pressure between the top and bottom of the Puy-de-Dôme[83] (and discovered that most of us live at the bottom of a sea of air). Medawar puts it this way:

The act performed to test an hypothesis may be called an 'experiment.' It is best to use the term in this simple and clean-cut way, rather than to follow common use in restricting its terms of reference to some sort of active messing-about with nature. A 'mere observation' may in this sense be an experiment.[84]

Applying experimental procedures thus defined is carried further in the next chapter.

Qualitative Changes, 1937–1939

It is surprising how quickly with a little thought, a scheme can be rejected.

Crick[1]

3.1 Ring Out the Old

I began my research career during 4 summers with the Ontario Fisheries Research Laboratory at Frank's Bay on Lake Nipissing, at the mouth of the French River. Thanks to the director, Professor W.J.K. Harkness, and to senior students, especially Fred Fry, I was lucky enough to get one of the finest introductions to field work that anyone could hope for.[2] Here also I began my interest in cycles, as Frank's Bay was the site of studies by Dunc MacLulich on snowshoe hares[3] and Doug Clarke on ruffed grouse.[4] Although I never worked up my limnological data, intended for a M.A., my efforts did much to counterbalance the dismal academic record of my earlier years at the University of Toronto. Transcripts were unknown in those days, and provided we did well in our final year we could, until then, sacrifice good marks in favor of freedom to enjoy the cultural and social amenities of a university campus, which I did in spades. Nowadays, poor marks in the first two years can hang like a sword of Damocles over a student's prospects. This is wrong. Cramming for marks is no way to get a rounded education; besides which, students must have time to discover their true interests. Although I had not then read Stephen Leacock, I now see that I subscribed to and acted according to his view that "a written examination is far from being a true criterion of capacity. It demands too much of mere memory, imitativeness, and the insidious willingness to absorb other people's ideas. Parrots and crows would do admirably in examinations. Indeed, the colleges are full of them."[5]

Having since climbed over to the other side of the fence, I take a sterner view of those who do poorly in examinations,[6] though knowing from sad experience that good marks are no guarantee that a student has the enterprise and imagination to do well in research.

After graduating in 1935 in Honours Biology, I went to Oxford, where I took the place of D.H.S. Davis for 1 year part-time and stayed 26 years. If I had graduated a year sooner, the job would not have been there; if I had graduated a year later, it would have been filled. Many another career has been shaped by similar accidents of time and place, also by the luck of being taught by good professors, as I was by A.F. Coventry, J.R. Dymond, and W.J.K. Harkness. All three were early admirers of Charles Elton and responsible for suggesting I make a temporary return to the land of my fathers. "Temporary" turned out to be longer than expected.

I spent the first half of my career doing science at Oxford University and the second half lecturing on the principles and history of science at the University of British Columbia. In this book I combine aspects of the two halves of my career, using the story of my research at Oxford as a framework on which to hang views about the process of science. The resulting smorgasbord of quotations is intended to allow students to choose those authorities that most appeal to them. Not everyone, for example, takes the same delight that I do in the writings of A.N. Whitehead.[7]

The most challenging of my teaching jobs was trying to persuade students in the humanities that science is not the cultural desert some of them think it is. Following James B. Conant,[8] I believe the strategy is to use historical examples to show the process of discovery rather than the end results. More recently, I've had some success using the Sherlock Holmes stories as a vehicle for introducing nonscientists to the principles of scientific inference (Section 11.2, note 58).

Having worked out mark-recapture techniques during my first year at Oxford,[9] I applied them to a population of voles at Lake Vyrnwy (Figure 3.1)[10] during a peak in 1936–1937 and decline in 1938 (area A, Figure 3.2). Also, I was lucky enough to find nearby populations that were at a peak in 1938 and declining in 1939 (area B). Unlike the large conifer plantations in which previous studies had been made, those around Lake Vyrnwy consisted of smaller units interspersed among sheep pasture, mature forests, and other unsuitable habitats. Also, the sheep that used to graze the land had been fenced out in different years, which meant that a succession of areas became suitable for voles, which may explain why some populations were out of phase with others.

This new study confirmed most of what was already known or suspected about vole populations, in particular that declines were not due to obvious changes in food supply, predation, disease, or reproduction. Nor were declines associated with dispersal; with one exception, no animals were caught in surrounding trap lines. Also, given the luxuriance and diversity of the plant communities in which the declines occurred, I found it implausible that food was deficient in quality.

The evidence against conventional explanations was admittedly too slight to convince those who were wedded to them. In such cases one has to choose between strengthening the contrary evidence or going ahead on a positive but new and risky line of one's own. Trying to discredit other people's fixed ideas

Figure 3.1. Lake Vyrnwy in September 1949. Young plantations, the only ones occupied by voles, were isolated from each other by sheep pasture, mature plantations, roads, and water.

is a losing battle (I've been on the losing side for 60 years), and except for the extinction of a few but powerful intellectual dinosaurs,[11] the only hope of progress is to provide a better alternative. So I hoped my line of thought would leave others dying on the vine. It's as well to get rid of such illusions at an early stage in one's career: faith in the relevance of starvation and predation continues to be a substitute for experiments (or to be protected in the face of evidence against them).

The evidence against starvation at Lake Vyrnwy was much the same as it had been at Newcastleton; evidence against predation was slightly stronger, as events were known in more detail. A uniform rate of decline would have been compatible with several hypotheses; but not only did the rate change abruptly, as at Newcastleton, but it affected males before females and was associated with low body weights and—on area A only—with failure of reproduction. Predators could not be blamed for these physiological effects, nor for the increased rate of decline. At the time of most rapid decrease owls had not stepped up their rates of predation, nor was there any reason to suspect other predators of having done so. Pellets collected from 10 selected fence posts fell off from 'many' in 1937 to four in January 1938, and only two thereafter during decrease and scarcity in 1938–1939.

Putting together these and previous results, I concluded that current ideas had directed us down the garden path. The argument went as follows: I had data for several populations assumed to be comparable. Let us assign letters to the various factors possibly relevant to the occurrence of a decline in numbers (E) and let +, −, and 0 mean 'present', 'absent', and 'not studied.' Suppose a = damaged food supplies; b = heavy predation; c = widespread disease; d = failure to breed; e = emigration f = bad weather; g, h, and i = unreplicated observations on parasites, competitors, and soil nutrients; and x = unsuspected factors. Now suppose that in population number 1 a decline (E+) was associated with damaged vegetation, heavy predation, and

Figure 3.2. Number of overwintered voles alive on area A during the peak of 1936–1937 and decline of 1937–1938. From Chitty (1952).

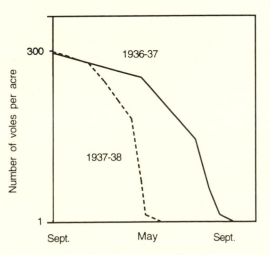

bad weather (a, b, f +) but not with disease, failure to breed, or emigration
(c, d, e −). We can then enter these data in line 1 of the table below, and
given two more instances of a decline in numbers (E+) and one in which
there was no decline (E−), we can complete the table as follows:

	Factors										
	a	b	c	d	e	f	g	h	i	x	E
Pop'n #											
1	+	+	−	−	−	+	0	0	0	0	+
2	−	−	+	+	−	+	0	0	0	0	+
3	+	+	+	0	−	−	+	+	−	0	+
4	+	+	+	−	−	+	−	−	−	0	−

This is the neat sort of table one can find in text-books on logic,[12] but in
the untidy world of population ecology anything this neat is a figment of the
imagination. And so it was with me. This table nevertheless epitomizes my
conclusions that because the variables a–d and f, in various combinations,
were only sometimes associated with E, they could be rejected as not necessary; and the combination of variables a, b, c, and f, could be rejected as not
sufficient, because it was present when E was absent. Variations in e, the
reproductive rate, were too slight, according to this scenario, to be relevant,
so one's next job would be to study the known variables g, h, and i and guess
the nature of the unknown variable x.

Inductive elimination does not, unfortunately, end in this clear-cut fashion.
First, there may be a 'complexity of causes,' which means that the supposedly
irrelevant factors, though not necessary *in toto,* may contain an element that
is necessary. (There may be a *disjunction* of factors.) As this last possibility
became one of the key assumptions in my new scheme of things, it's worth
explaining it in more familiar terms. Wine that maketh glad the heart of man
can can be made not only from grapes, but also from blackberries, elderberr-
ies, pears, and other fruits. No particular kind of fruit is necessary for this
state of gladness; but all provide at least two common elements that may be
necessary, namely, water and alcohol. Of the two, we infer from a wider
body of knowledge that alcohol is the relevant factor. The analogy with the
vole problem is that bacterial disease, viruses, parasites, predation, bad
weather, etc. may have a common element that must be present for a decline
to occur. In other words, no predictable mortality factor need be present in
its entirety, but each may include some necessary element.

Second, there may be a *conjunction* of so-called causes of death. Though
neither food shortage alone nor predation, for example, may be sufficient to
produce a decline, the two together may do the trick.[13] Plausible solutions of
this kind cannot be dismissed except experimentally, however suspicious one
should be of an extra hypothesis introduced to make up for the deficiency of
the first. Moreover, there is no end to the process. If a 2-fold combination

should fail, others can always be found after an event such as a decline in numbers, for example, dispersal, together with aggression, low reproductive success, poor quality herbage, disease, and predation. 'Arm-waving' of this sort is not unknown to writers of text-books,[14] who perhaps feel obliged to present their readers with irrefutable solutions.

We now turn to further evidence that brought about these changes of thought.

3.2 Ring in Maternal Stress

Whatever is written will remain permanently in the literature and one's scientific reputation can be damaged by publishing something that is later proved incorrect. Generally speaking, it is a safe policy to give a faithful record of the results obtained and to suggest only cautiously the interpretation, distinguishing clearly between facts and interpretation. Premature publication of work that could not be substantiated has at times spoilt the reputation of promising scientists.

Beveridge[15]

Population changes on my original marking area at Lake Vyrnwy[16] consisted of (1) a gradual decline during the worst of the nonbreeding winter of 1936–1937; (2) a sharp but limited decline in spring 1937; (3) little or no recruitment during the summer in spite of a high pregnancy rate; (4) increase by autumn to a density similar to that of autumn 1937; (5) a gradual decline in early winter; (6) a severe decline in late winter and spring of 1938, in which only 2% of the overwintered animals were still alive in June compared with 20% in June 1937 (Figure 3.2); (7) no breeding that spring, and scarcity throughout 1938 and 1939 in spite of an apparently favorable environment.

In four respects, results from this study differed from earlier ones. First, when taken into the lab in 1938, voles survived for a long time after the decline of those in the field, which threw doubt on the interpretation of previous poor survival rates in the lab. Second, some young females matured in their first year, though towards the end of the peak season only. These results were contrary to the earlier inference that young of the year fail to mature, the authors of this claim having had the misfortune to generalize from data for 2 years only. They were apparently right about the males, however, as few, if any, matured during their first summer. Indeed, most young of both sexes disappeared *in situ* during the height of the 1937 breeding season, none having been captured in nearby trap lines. Survival improved later that summer, but as the males remained immature, their sisters presumably bred with the few overwintered males still alive. Third, voles began breeding later and stopped sooner than expected. Fourth, and finally, as already explained, disease did not explain why the voles declined in numbers.

One new discovery was that body weights were lower during the decline of 1938 than they had been at the peak; and thanks to having added several out-of-phase populations to that year's study, this apparently poor condition could not be blamed on the weather. For, at the same time that voles with low body weights were disappearing from the original area (A), others of similar age or younger[17] on nearby areas (B) had high body weights and were at peak densities. Given these data we can compare weights between years on area A and between areas A and B in 1938 (Figure 3.3). Though not independent, the two ways provide useful comparisons. Until several years later it seemed obvious that low body weight was a pathological condition associated with whatever was killing off the population.

The problem now was to find the unknown reason for the difference in survival rates on area A between 1937 and 1938. As numbers and environmental conditions had been similar in two successive winters, the only difference I could come up with was in the animals' origins. For the overwintered animals of 1937 had been born in an expanding population, and those of 1938 in a stationary population in which the early-born young had suffered a high death rate. I inferred that these young had been killed by the adults and that their disappearance was a symptom of competition for space at peak densities. The later-born young, by contrast, survived without suffering direct contact with the adults, most of whom soon died or stopped breeding. Also, the body weights of these overwintering young were normal at the start of

Figure 3.3. Distribution of body weights of male voles in spring during peaks and decline. From Chitty (1952: Figure 2).

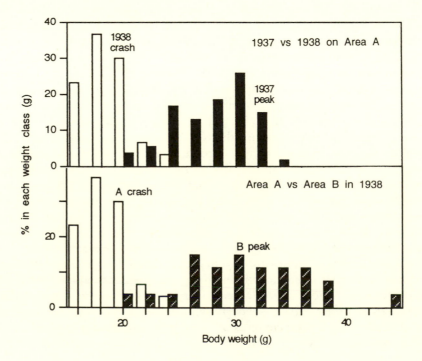

winter. So I had to think up a way by which a change in the parental generation could affect the later survival of their offspring.

The idea that natural selection might explain the difference between generations was unacceptable then and for some years to come. In 1950, for example, at a meeting of the Society for Experimental Biology, John Clarke and I
explained what we were up to and were asked if we had considered genetic
change as a possible explanation. John replied as follows: "My genetic colleagues at Oxford believe this requires too rapid a genetic change, beyond
the speed of normal selection." Thereupon, J.B.S. Haldane rose majestically,
turned towards John and said: "Your Oxford colleagues are quite correct."
This incident effectively closed our minds to natural selection, and for the
next 7 years I lacked the courage even to mention its possible role. John took
even longer: it took him 38 years to show that our advisors had been gazing
into a clouded crystal ball.

As the only alternative seemed to be a maternal effect, I reasoned that
'competition for space' had disrupted the mothers' hormone balance in some
way that affected the viability of their young. I also inferred that the last
litters were even less viable. I'm now sure I misinterpreted this evidence, and
that the worse survival of the later young was due to their being born into a
more crowded population. At the time, however, the idea of some abnormal
condition being transmitted through more than one generation was consistent
with the evidence. It also explained the evidence for snowshoe hares if one
allowed for their greater life span and stretched one's imagination to allow
the ills of the mothers to be visited upon the children even until the third and
fourth generations. The vole and snowshoe hare cycles seemed to have too
much in common to need separate explanations, a belief that has survived
my original faith in maternal effects.

A qualitative difference between individuals in different phases also offered
a solution to the problem of synchrony. Exceptions show that no regular
climatic cycle can be directly responsible for the phenomenon; but the tendency for independent populations to be in phase must be due to some coor-
dinating agent. How might it work? The effect of any such agent can be
measured only through the response of the animals. So if they became less
viable as the cycle progressed, the *effects* of weather would become more
severe even if the *weather itself* remained the same. Populations in different
phases would then respond differently, and the rate at which they came back
into synchrony would depend on the variability of the weather, the regional
extent of the variations, and the initial density of the populations.[18]

One of the difficulties about accepting such a role for weather was to
account for the way in which different species, among them snowshoe hares
and grouse, so often fluctuate in synchrony. I suggested that weather would
synchronize such cycles if and only if it were powerful enough to "reduce the
effect of specific differences in population parameters." But if so, one would
expect to find cycles varying in length within species "whose range included
a wide variety of climatic conditions." At the time there was no evidence to
support this implication; there is now.

These thoughts had not all matured before the outbreak of war and were not submitted for publication until 1951. They were received with pursed lips by the new editor of the *Journal of Animal Ecology,* by two of my colleagues, and by miscellaneous critics, and indeed had to be modified later. Part of the trouble was the absence of experimental or theoretical evidence to support a long-lasting maternal effect. Second, these ideas conflicted with the establishment's faith in the theory of density dependence, according to which the relevant changes are in food, predators, parasites, and disease, all of which I considered irrelevant to the occurrence (but not the rate) of a cyclic decline. This theory also claims that the action of weather is independent of population density, a claim for which there is still no shred of evidence. According to my beliefs, conventional wisdom had failed to identify the variables relevant to cyclic populations and might be equally useless for noncyclic populations, not only of voles but of other organisms.

As these views were new, heretical, and difficult to test, I felt nervous about publishing them, and until provoked by their hostile reception had read too little in the philosophy of science to be certain about the role of speculation. It was therefore reassuring to find Darwin saying that "false views if supported by some evidence, do little harm, for everyone takes a salutary pleasure in proving their falseness."[19] Providing this kind of pleasure was not what I had in mind, however.

3.3 Age Structure

When entering on new ground we must not be afraid to express
even risky ideas so as to stimulate research in all directions.
Bernard[20]

P.H. ('George') Leslie joined the Bureau in 1935. None us knew why he was called George. Besides helping us all with his statistical wizardry, he put Ranson's vole data into actuarial form and made many other original contributions. Unlike some mathematicians, he had firsthand knowledge of the real world of the field worker, came on many trips to Lake Vyrnwy, and more clearly than Charles realized that bright ideas should be taken lightly until tested experimentally. Noting how seldom population theories were disciplined that way, he had reservations about them that are too seldom recognized by today's theorists. "It is an amusing, and a far from unprofitable speculation," he wrote,[21] "to consider the way in which theoretical physics might have developed if only an occasional experiment had been done." He also emphasized the limitations of his own models. He spent many profitable hours with Tom Park, discussing, among more learned topics, the stories of Somerset Maugham and Conan Doyle. They much admired both authors and thought their importance underestimated.

One of the prewar activities at the Bureau was maintaining a lab stock of voles. The stock was going strong when I arrived in 1935, but Charles told me they had had trouble getting it started. Apparently, the tide had turned with the arrival of a batch of voles from Corris. It is intriguing to speculate that the early failures were due to the voles having come from low-phase or declining populations instead of from the apparently noncyclic population at Corris.

In establishing a domestic stock, one is apt to try all kinds of tricks with diet, lighting, space, and social conditions, and to attribute eventual success to the last of these changes. Richard Ranson, who looked after the domesti-cated stocks, revealed his tongue-in-cheek secret for breeding wood mice: after months of failure, he got them started by providing an inverted flower pot.

Richard designed his breeding program for studying longevity and fecun-dity, and George Leslie used the data to produce a life table.[22] Like all life tables, this showed that the older the animal, the lower its expectation of life. This was no surprise; but George applied the idea to the vole cycle and sug-gested that once a population had become stationary through reduced breed-ing its members would be older than usual and thus likely to die from quite small environmental disturbances. Such a population, he suggested, was bound to fluctuate now and then.

At Lake Vyrnwy, however, the voles behaved in the opposite way from those in George's model, as birth rates remained high in the year before the crash. Instead, the stationary state of the summer population of 1937 was due to poor survival of the young.[23] At the end of the season, however, the young survived well, and as most adults died soon afterwards, the population that declined in spring 1938 consisted of young animals that had spent the winter in a nonreproductive state. By contrast, the autumn population of the previous year would have included animals born throughout the phase of increase, not just the final litters. Or so I thought. Accordingly, I rejected George's explanation for the decline.

I would not now use the same argument to reject George's hypothesis. At the time, the increase phase was a closed book, and I could only assume that litters born throughout the increase summer would have survived to form part of the overwintering population. If so, the animals would have been older in the peak population that did not decline, and younger in the popula-tion that did. Only later did I realize that breeding in the peak year (1937) had stopped earlier than it does in an increase phase. So George was probably right after all; but, in fact, age in months is meaningless among animals in-hibited from reaching sexual maturity; 'old age' is postponed in such animals. As well, we now know that declines are just as severe when they follow a phase of increase as when they follow an intervening peak year.[24]

The argument I now prefer is that the acceleration in death rate in late winter and spring 1938 was too rapid to be attributed to the onset of senes-cence. The animals had survived the worst of the winter far better than one would expect of an aging population. They then survived far worse, especially the males. I later rejected my own explanation on similar grounds, namely, that viability was also unlikely to change so abruptly.[25] Thus, a higher-than-usual age structure seemed neither necessary nor sufficient to explain a decline. Nevertheless, a sophisticated version of George's hypothesis surfaced some 54 years later, accompanied by the odd statement that "shifts in age structure" was a solution to the cycle problem that had been ignored until then.[26] The author was generous enough, however, to give Leslie and Ranson a two-word eulogy as "notable exceptions."

I must confess that, in the belief that the voles were younger than usual during a decline—instead of being older as predicted by George's model—I paid no further attention to this explanation. Only later did I realize that something more general than a change in age structure was implicit in the idea. The distinguishing feature of a life table is that, while showing the probability of death, it does not specify its causes. Insurance companies know the rate at which their clients are likely to die, but not what each will die of. They are concerned with the probability of dying and how it differs between one set of conditions or clients and another. The implication for mammalogists is that a life table merely predicts the rate at which animals are going to die regardless of what finishes them off. In the lab, for example, a difference in humidity, temperature, lighting, or diet will explain a *difference* in the shape of two life curves but will not tell us what the animals die of. All the animals dying in Richard's cages were sent to Dr. Wells, and the reports always came back N.A.D.—nothing abnormal discovered.

Figure 3.4 illustrates George's point that the greater the average age of a cohort, the faster it will die out, though living under the same conditions as a younger cohort. Curve *a* is plotted from the smoothed life table for young male and female voles living in Richard's basement in the old zoology build-

Figure 3.4. Percentage alive: (a) young prewar voles; (b) the same when one year old; and (c) young postwar voles. From Leslie and Ranson (1940) and Leslie et al. (1955).

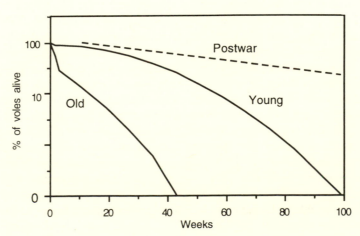

ing. Curve *b* is plotted from the same data, but only for animals that were already 1 year old. The latter become extinct about 60 weeks sooner than the cohort of newly weaned animals. Curve *c* is plotted from data for voles living under a different system of husbandry after the war. The difference in slope between curves *a* and *c* is probably accounted for by an improved diet, perhaps also by a difference in genotype; but in neither case do we need to know what the animals died of.

George also calculated the intrinsic rate of increase of a vole population subject to the birth and death rates of the lab stock. Such a population would increase 10-fold in 6 months, which is probably below the rate my postwar stock might have achieved. Knowing that all populations are able to increase geometrically, I took no special interest in this phase and concentrated on the decline. It turned out, however, that there's more to the phase of increase than meets the eye.

3.4 Shock Disease

An apparently arbitrary element, compounded of personal and historical accident, is always a formative ingredient of the beliefs espoused by a given scientific community at a given time.

Kuhn[27]

At the same time that the epidemic hypothesis was being rejected at Oxford, it was getting similar treatment at the University of Minnesota, where several workers were studying the '10-year' cycle in snowshoe hares.[28] They, too, found no evidence that an infectious disease could explain a decline in numbers. Instead, they believed that the hares were suffering from a metabolic disturbance called 'shock disease' and concluded that it was "responsible for the die-off of the snowshoe hare throughout its range in North America." Evidence in favor of this hypothesis was as follows:[29] (a) large numbers of hares died in the wild during 1936; (b) a higher proportion of hares died in the traps in 1936 than in the 2 previous years; (c) hares kept together indoors in the spring of 1936 died at a high rate with characteristic terminal symptoms; (d) a population breeding in an outdoor enclosure declined in numbers in 1936; (e) the symptoms of death in nature, in the traps, and in captivity were alike, which implies that the primary cause of death was the same in all instances.

The deaths in captivity were spectacular. On arrival from the field, animals were put into a room with 560 sq ft floor space. One shipment consisted of 48 hares. Two days later a second shipment of 90 was added to the survivors. Average survival time was 4 days, which was accepted as evidence that the animals had been abnormal before capture. Some died abruptly, others after symptoms lasting several hours, including severe convulsions. A frequent symptom post mortem was low glycogen content of the liver and abnormal levels of blood sugar: usually too low but sometimes too high.

As with the vole work, nothing in the external environment could be cor-

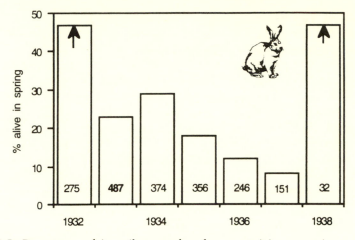

Figure 3.5. Percentage of juvenile snowshoe hares surviving overwinter on a 1.2-square-mile area. Numbers show the size of the breeding population into which they were born; peak year in heavy type. Arrows show minimum percentage alive. From Green and Evans (1940).

related with the high death rate, and coming at a time when no alternative hypothesis had been established, 'shock disease' seemed a likely explanation for cyclic declines. In fact, almost everyone, myself included, climbed onto this bandwagon, which rolled on unchecked for almost 20 years. The main conclusion was that the decline was due to changes in juvenile survival, which continued to get worse in spite of decreasing population density (Figure 3.5). The survival rate of adults and yearlings was not unusually poor except at the lowest point of the cycle.

Having been published partly on the eve of of World War II and partly after it began, this new hypothesis was one of our less pressing concerns during 1939–1946; so the evidence lay ticking like a time bomb until work on cycles was resumed. Then both John Christian[30] and I accepted it as the answer to a mouser's prayer, a view that had a profound influence on the course of later work, as it encouraged many of us to study physiology and behavior instead of food supplies and predation. Although shock disease turned out to be a mere artifact, it shows how falsity of a premise (the existence of shock disease) does not imply falsity of a conclusion (one should study physiology and behavior), for "wrong reasoning sometimes lands poor mortals in right conclusions."[31]

This section completes the survey of the state of the art at the outbreak of World War II. In addition, we had been conducting two enquiries, one on lemmings and other arctic animals and one on snowshoe hares. Helen kept these enquiries going, but for the next 6 years we called a halt to the rest of the work.

3.5 The First Ten Years

Just as important as having ideas is getting rid of them.

Crick[32]

The state of the art at the end of the first 10 years' work is given in *Voles, Mice and Lemmings:*

> The picture at present is of predators too scarce and disease too erratic to explain the persistent cycles . . . Food, except possibly in the critical winter months of a peak abundance, is almost unlimited . . . Without at present putting the matter any higher than this, we can see that there are inherent properties of the population dynamics of voles that may eventually explain their cycle as a self-contained system that is not so much dependent on other animals like predators and parasites as we at first supposed . . . Voles drop off at all times of life, though not at the same rates. And these are not 'ecological' deaths; few of them probably are 'parasitological' deaths. We hardly know what process is at work, and for want of a better term we may call it 'wear and tear.' This has the suggestion of internal break-down in the physiological organization. We might almost say that the process of senescence begins at birth. This basic mortality presumably goes on in all vole populations and therefore might be expected to play some part in the cycle.[33]

The concept of 'wear and tear' meant that death rates—from predation, to take one example—now had to be looked at differently: "a vole that is, say, killed by a kestrel, might have died *in any case* next day or next week, in its ordained place in the life-curve."

Most writers on cycles refer to *Voles, Mice and Lemmings;* fewer have studied it; fewer still have learned its lessons. Perhaps they have taken too literally the passage in which Charles wrote that he "would put as much emphasis on the methods of approach to the subject as on the facts already learned." He did not mean that future workers should disregard the so-called facts[34] and start back at square one. He surely meant that they should adopt his methods to advance the subject from the ground so painfully gained. He had expected that when we had understood the system we would have learned "new principles that will apply to many other populations of animals."

Workers on other populations have been slow to accept this vision, partly because of the strength of their own preconceptions, partly through relegating cyclic populations to a unique class and seeing no analogies with populations of other species; but also, I suggest, because even mammalogists have ignored the evidence in *Voles, Mice and Lemmings* against conventional views about the role of food shortage, predation, and disease. The evidence convinced those of us who had helped collect it, but we had been unable to do more than suggest an alternative approach, not reach a solution. And until there's a convincing substitute, old ideas never die and seldom fade away.[35] The gist of the new line of attack was to study factors affecting the probability of dying rather than the mortality factors themselves. That, at least, was how I interpreted the evidence. It turned out later that Charles was unconvinced by his own logic.

Besides stressing the need to look in other directions, Charles satirized previous explanations for declines.

> Voles multiply. Destruction reigns. There is dismay, followed by outcry, and demands to Authority. Authority remembers its experts or appoints some: they ought to know. The experts advise a Cure. The Cure can be almost anything: golden mice, holy water from Mecca, a Government Commission, a culture of bacteria, poison, prayers denunciatory or tactful, a new god, a trap, a Pied Piper. The Cures have only one thing in common: with a little patience they always work. They have never been known entirely to fail. Likewise they have never been known to prevent the next outbreak. For the cycle of abundance and scarcity has a rhythm of its own, and the Cures are applied just when the plague of voles is going to abate through its own loss of momentum.[36]

This passage has a significance well beyond its criticism of the Pied Piper and other cures. It's a powerful warning against committing the fallacy of *post hoc, ergo propter hoc*. In fact, it was this fallacy in the papers on *Toxoplasma* (one of them co-authored by Charles himself) that first showed me how easily one can fall into this trap. But beholding this mote in someone else's eye did not prevent me, some years later, from getting a beam in my own eye.

Without any extra mortality imposed on them, certain populations are going to decline anyway. We must therefore be skeptical of mere *a posteriori* claims for the success of biological and other control measures. For example, evidence that an insect pest has been controlled is unacceptable without evidence that the outbreak persisted in areas to which the virus, parasite, or parasitoid had not yet spread. The need for controls is a theme I shall harp on throughout this narrative.

Charles could point to an impressive advance in knowledge and methodology but not to an immediate likelihood of solving the cycle problem. Vole dynamics, he thought, might be completely understood after a further 10 or 20 years work, an optimistic prediction that has long been falsified. Luckily, hope springs eternal in the ecological breast, so that in 1988 we find Lidicker saying "we are actually quite close to an adequate understanding of microtine cycles."[37] We should none of us hold our breath.

4

Wartime Rat and Mouse Control, 1939–1946

Pedants sneer at an education which is useful. But if education is not useful, what is it?. . . . It is useful, because understanding is useful.

Whitehead[1]

You know, this applied science is just as interesting as pure science, and what's more it's a damn sight more difficult.

Hardy[2]

When war broke out in September 1939 we switched from pure to applied zoology, and thanks to Charles's foresight the transition was fast and smooth. Our mandate now was to help protect Britain's food supplies from rodent pests. For every ton of food that could be saved, a ton of other goods could be imported. Instead of roaming the hills of Wales and Scotland we now spent our lives at garbage dumps, down sewers, on a pig farm by the tracks, below stairs in restaurants, and anywhere we could study Norway rats or house mice. Luckily, our work also took us into the country (Figure 4.1) at a time when few people were able to travel. Several other workers joined the Bureau, and research was extended to London and abroad.

Pure and applied science differ mainly in aims, not methods. The pure scientist studies God's works the theologian studies His words, but usually reaches a different conclusion on a number of thorny topics. The applied scientist recognizes that though man cannot live by bread alone, he can't do without it; he therefore applies existing knowledge or adds to it as required to solve practical problems. Unlike the pure scientist, he is not judged by his contribution to knowledge or the hope of its being useful at some distant date. He must limit his curiosity to what he considers relevant to a specific problem. "Technology is of the earth, earthy . . . it is under an obligation to deliver the goods."[3] To do so may involve starting expensive projects on incomplete evidence; time is of the essence, and the risk of disaster must be faced. Charles described our position as follows:

> The usual way of arriving at a balance of probability about the truth of a piece of research is through its publication and subsequent criticism and independent appraisal by other workers. For an ecological idea or method, 10 years would

Figure 4.1. How not to approach a farmer for permission to study his rats. Cartoon kindly drawn for me by the late Doris Gully.

be a short time in which to go through this process properly. During the war there was not time for this. . . .[4]

Such difficulties make applied research more difficult than pure research. As well, one is less likely to find inspired solutions when marching down a route mapped out by a committee, or by bureaucrats holding the pursestrings, than when meandering along a path of one's own choosing. This is a difficult truth for governments to grasp, as few of its decision makers have any understanding of science. One of the world's longest and best population studies[5] was canceled at its most fruitful stage, as it was said to be giving no practical results. A comment by Medawar is appropriate:

> In these days of cost-consciousness, when the funding of research is being administered as if scientific research were a branch of the retail trade, and when "pure science" and those who practise it are coming under an increasingly cynical scrutiny, it is as well to remember the definition that Oscar Wilde puts into the mouth of Lord Darlington in *Lady Windermere's Fan:* a cynic is "a man who knows the price of everything and the value of nothing."[6]

Once there's a possible solution to a problem it must be put into practice, at first on a small scale known as operational research. One has to find out if other people can handle the techniques, what compromises to make, and what training to provide. Then, if all goes well, the technology can be used on a large scale. This, at least, was how our own studies advanced from individual tests, through training sessions and small-scale operations, to a highly successful national campaign. The key to our success was prebaiting, which is a system the Borgias would have approved of for gaining the victims' confidence before slipping poison into their repast. So successful was this system that we were soon unable to find local rat populations except in sewers and on a small pig farm (Figure 4.2), where we bribed the owner to refrain from harassing his flourishing rat colony. I spent many evenings rat watching in one of the sheds, sometimes distracted by the roar of German bombers on their way to Birmingham or Coventry.

Control of Rats and Mice gives the story of this research and its application to conditions in wartime Britain.[7] We were lucky in being given a job to which we could so easily adapt our experience with animal populations. Those who doubt the value of pure research may take comfort from the thought that ours came in handy. One of my prouder moments was listening to an encomium delivered by a London sewerman about an apparatus he didn't know I'd invented (Figure 4.3). It transported rat baits from street to sewer level and saved this rather portly gentleman from climbing up and down a ladder. The device was the reverse of the Eckman dredge I'd learned to use in my undergraduate days on Lake Nipissing—another example of an unexpected pay-off from pure research.

Even if we'd been assigned some other job, it's likely that our research experience would have been useful in the narrow pragmatic sense. Biology is an excellent intellectual discipline. Its universe is more chaotic than that of physics or chemistry and teaches its practitioners confidence to act on judgments they can seldom justify until later (if then). Many of Pasteur's greatest

Figure 4.2. Science in action at a piggery. See Chitty (1954b) for a photo of the piggery and Chitty and Shorten (1946) for a photo of one of its marked rats. Cartoon by Doris Gully.

Figure 4.3. Baiting sewer rats from street level.

discoveries "were published in the form of short preliminary notes long before experimental evidence was available to substantiate them."[8]

Because biologists are accustomed to dealing with uncontrollable variables, they were highly regarded in the Royal Air Force, or so David Lack
told me. He himself discovered that the "angels" that appeared on radar
screens, and looked like some secret weapon, were due to the dispersal of
starlings from their nighttime roosts.[9]

We began winding down applied research in 1946, but I was unable to
return to full-time work on cycles until I'd written up my own rat work,
edited other people's, and proofread the two volumes, which were published

in 1954. It's possible that this task delayed our fundamental work, much of which was never written up; but it's also possible that a lapse of time is needed for ideas, like eggs, to hatch. In any case, compared with contemporaries at Oxford and elsewhere, I was lucky to have suffered so slight an interruption to my career. I was also lucky enough to be sent on a coast-to-coast trip across the United States and back, meeting others who had worked on rodent control.

One port of call was Baltimore, where I met John Calhoun and David E. Davis and got an advance look at work in progress. They were later joined by John Christian. Unlike us, these workers and John Emlen before them used their rat studies to develop fundamental ideas about populations in general.[10] Using rats as their experimental animal, they too concluded that rodent populations are able to regulate their own density. They were impressed by the way populations reduced by traps or poisons returned to a characteristic level below that set by food or cover. Population density did indeed fall when resources were reduced but leveled out or fell to near zero while food and cover were still plentiful. The inference that population density was set by behavior was confirmed by adding strange rats to a population or substituting strange rats for those removed. Populations density either fell or stopped increasing, presumably due to the resulting aggressive behavior. In 1949, John Calhoun published his evidence that "a physiological and psychological disturbance in socially inhibited individuals . . . might have a deleterious effect on the progeny either through poor foetal nutrition or from breakdown of maternal instincts."[11] This was encouraging; and though our interests soon diverged,[12] we kept up a friendly correspondence for many years, though continuing to find the others' arguments less persuasive than our own.

Replication, 1946–1951

Reading the 'experimental methods' section of most scientific papers is more boring than almost anything I know.

Crick[1]

5.1 Postwar Methods

Helen Marie Stevens graduated in premedical science at the University of Toronto and for her M.A. switched to the Department of Biology, where she studied the pituitary body in the cat.[2] She also became a teaching assistant in my third-year class, but as she gave me the wrong answer to a question, I took no further interest in her until 1 year later, when I needed someone living within walking distance to take to a dance. The lucky fact of my being unwilling to pay for a taxi has shaped much of my scientific as well as personal life. In 1936, after finishing her M.A., Helen joined me in Oxford, where we got married. She first worked with Doug Middleton on partridges, then on the Canadian Arctic Wild Life and Snowshoe Rabbit Enquiries, then part-time on voles. It may seem that an anatomical study of the cat's pituitary was a poor preparation for ecological research. The function of a postgraduate degree, however, is to prepare students to tackle problems of any kind, not just those on which they cut their teeth.

Like other women who combine research with raising a family, Helen had to find a balance between competing claims.[3] Her difficulties were further compounded by being married to a compulsive research worker, who took it as his God-given right to put work before domestic duties. Unlike me, she could not neglect the children to work long uninterrupted hours or fail to keep the house running smoothly. It was always she that took time off when the children were sick and she that subordinated her research career to mine. Fortunately, being a scientist herself, she understood the difficulty of getting anywhere in science unless, when on a roll, one is free to work long and irregular hours. "Wives, children, houses, regular hours are the bane of committed laboratory research," says Watson.[4] Substitute husbands in this list and you have a fair idea of the difficulties faced by women in science, particularly if they are field workers. In having as considerate a boss as Charles, Helen was luckier than many women, as her job was compati-

ble with flexible hours and time off for parturition, lactation, shopping, cooking, house cleaning, and other functions that husbands are obliged to delegate or feel themselves unqualified to perform. Helen did a lot of work for which she got little credit, and without her selfless support I would have had a far less happy professional life. Watson notwithstanding, many a wife is the unsung heroine of her husband's career. Helen's death in 1987 was so shattering that research no longer seemed important, and by the time I'd regained my composure, a whole filing case, containing notebooks for the previous 10 years, had disappeared. Fires, floods, loss, thefts, sabotage,[5] and other catastrophes are among the perils that all of us, graduate students especially, should try to guard against. Thanks to computers, one can now back up data more easily than in the antediluvian days of hand-written records.

In the 1930s, few male academics were faced with the dilemma of being wedded to an equally or better qualified wife. Life is more complicated now that both spouses want jobs and may be unable to land them in the same city, let alone in the same university. One of the pair may have to tag along with whomever gets the best offer. Thanks to Charles, Helen and I were spared the difficulties facing so many of today's marriages between scientists.[6]

On getting back to Lake Vyrnwy in 1946 we needed to replicate the prewar observations. Four aspects had to be checked out: body weights and reproduction (from snap-trapping); survival in the field (from mark-recapture); food supply (from comparative observations); and survival in the lab (from life tables). Responsibility for these studies now rested with Helen and me (Figures 5.1, 5.2). We first reviewed our techniques.

To begin with, we did not know what phase of the cycle we had come in on, so went back to using the prewar indices of abundance. As we spent several days on each visit, we seldom used the quick but boring method of trace index. Instead, we set several lines of snap-traps, which gave us a rough index of abundance and satisfied the need for corpses.

On each visit we set traps 5 or 10 yd apart in several lines spaced out over the whole estate and ran them for 1 to 4 days. New lines ran parallel to and 25 yd away from earlier lines, none being retrapped until the voles had had a year to recover. Being obliged to trap different ground on each visit meant that comparisons between months were confounded with effects due to differences in habitat. Also, as other duties prevented us from traveling at the same time each year, the index of spring abundance was more variable than it might have been. A difference of a couple of weeks in the onset of a decline could make a great difference. Although rough and ready, the trap index enabled us to follow the direction, though not the magnitude, of changes in abundance as well as differences in population density between areas.

By 1950 we knew that nonrandom sampling of marked and unmarked voles prevented us from measuring total numbers, though not from measuring changes in survival. As our main aim was getting samples of animals for

Figure 5.1. Strengthening the pair bond. Picking up snap-traps at Lake Vyrnwy. November 1948. Photo by Denys Kempson.

post mortem examination, we therefore reduced our marking studies to quick ways of measuring changes in survival. A simple but effective system relied on the tendency of voles to stay in one spot until they died. In the peak summer of 1952 we marked animals at a dozen trapping stations spaced well apart over a 6-mile transect.[7] Each had a surplus of traps around a single point and was visited each month. Of 20 adults marked in May, 7 were still alive 4 months later, a minimum survival rate (P*) of about 0.9 per 2 weeks. This transect system sampled a diverse area with less expenditure of time than in previous studies. It also meant that errors due to keeping mothers away from their young would be thinly spread over a large area. The disadvantage was that, with a linear design, minimum survival rates were probably reduced still further by dispersal. In the following two winters we therefore adopted a modified system of trapping at 15 points in each of three grids, 25 yd × 15 yd, spaced well apart in different habitats.

A third variation on this system, used by Janet Newson and me in 1956–1958, was to set about 50 live-traps spaced 15 yd apart in a wide circle. It had the advantages of the transect system plus the chance that if voles moved from their original places, some would do so across a chord of the circle and provide data on movements (which grid-trapping tends to restrict). Also, the 15-yd spacing meant that few traps competed for the same voles. By this method we were unable to estimate population density, which in any case was tricky because of the nonrandom sampling just mentioned, but we could easily estimate minimum survival rates over a transect of almost half a mile

Figure 5.2. Charles Elton, Thomas Park, and Mick Southern at Lake Vyrnwy in November 1948: checking up on the author. Photo by Denys Kempson.

of mixed habitat. We were thus able to achieve our objective of looking for correlations between survival and other variables, such as vegetation.

The problem of nonrandom trappability had surfaced when I got around to writing up my D.Phil. thesis. In the first draft I stated that marked and unmarked animals, as far as was known, were equally at risk of capture. It's a good idea to be explicit about one's basic assumptions; it's often a shock to find no evidence for them. So it was in the present case, and a few simple analyses showed that new animals were apparently reluctant to enter the live-traps. A possible reason for their behavior had been planted in my subconscious by our wartime work on rats. We had learned about prebaiting, which was essential for the success of quick-acting poisons. We had further extended this knowledge by discovering rats' so-called new object reaction to almost anything new in their environment, unfamiliar foods included. The factor that enabled rats to avoid a trap was not human odor, as tradition maintained. In the course of rat watching we saw them avoid such harmless objects as a piece of wood handled with rubber gloves and placed between them and their food. Several days later, after the wood was removed, the rats continued to use their well-marked runway around the place it had occupied.

So we now thought we understood some of our difficulties with live-trapping voles. Much of my life at Oxford and some of Helen's had been spent in frustration at being unable to design an improved live-trap. We would design a trap, drive a hundred miles or more, set out a line of snap-traps and a parallel line of our latest brainwave in live-traps. The next morning the catches would be in the ratio of perhaps 10 to 1 in favor of snap-traps. So back we'd go to Oxford, modify the design, and repeat steps two to five. If we'd had patience enough to leave the traps prebaited for a few days we would have found that almost any design would work. So there had been no need for the complicated double-entrance trap that Helen, Charles, and Doug had helped me invent.[8] The reason it seemed to work so well was that I had tested it on marked animals. Having once enjoyed its food and shelter, voles were happy to be trapped time and time again. Unmarked animals, however, took one to several days to get over their trap shyness. Or so we thought at first; but there was more to it than that, as we learned a few years later.

Having recognized the need for prebaiting, we now made better progress on designing a new trap. We would try out a design in the field and bring back something that might have been invented by Rube Goldberg or Heath Robinson, which we would present to Denys Kempson, our superb technician (known as 'DK' to avoid confusion with a similar-sounding name around the Bureau). DK would turn our amateur effort into something resembling a Leica camera, which we would take back to the field. The result of this collaboration was the Longworth Trap,[9] which brought at least a small part of the world beating a path to our door. If we'd anticipated the popularity of this trap, we might have taken out a patent.

If mere unfamiliarity made unmarked voles shy of a new object, a few days prebaiting would presumably make them as trappable as the marked animals. So in our first postwar live-trapping we began by propping the door open for a few days. But the trouble still persisted. Mark-recapture statistics[10] implied that new animals were entering the population throughout the winter. This could not be, however, as we had chosen an isolated area and knew when breeding stopped. Our trapping was still biased. New animals were indeed being recruited, but not into the population as a whole, only into the marked population. Perhaps certain animals were keeping out of the way of those that entered first; but for whatever reason, their behavior made it difficult to estimate total numbers of field voles, though bank voles were less reluctant to enter the traps. Using various tricks, we could still get a rough idea of total numbers in the nonbreeding season, and George Leslie invented ways of dealing with the data at all times of year.

We also used prebaiting to improve the efficiency of trapping live animals to take to Oxford. To avoid having traps sitting idle, we made dummy traps that nested inside each other like paper cups and so were light and easy to carry and assemble. We set long lines of prebaited dummies and left them alone for the first 2–3 days. We then put real traps only where the dummies had been entered. This method was particularly efficient when voles were scarce and we had to make the most of every trap. Using the same principle during a cyclic low, Boonstra et al.[11] tripled the number of lemmings per trap compared with the numbers they caught in grid-trapping.

5.2 Postwar Body Weights

There isn't such a thing as a hard fact when you're trying to discover something. It's only afterwards that the facts become hard.
Crick[12]

In my first study at Lake Vyrnwy, body weights in spring were high in peak populations and low in the decline, an association I interpreted as being a change from a normal to an abnormal condition. In our immediate postwar work, however, we discovered that high body weights were the exception and that low body weights might persist, in spring, for 3 of the 4 years of a typical cycle.[13] Variations cropped up later on: if breeding began early in the increase phase, weights were as high as in a peak year, and later still it turned out that under some circumstances heavy animals might still be present during a decline. We also found a relationship between the phase of the cycle and the distribution of body weights in June (Figure 5.3). These data are plotted to show the proportion in each weight class, with the lowest at the bottom and the highest at the top. In 1947 and 1951, for example, when the population was increasing, all age groups were represented; but in 1949, when the population was declining, the really heavy animals were missing and few young had been recruited. In the peak year, 1948, there was a gap between the heavy overwintered males and immature young, as was to be

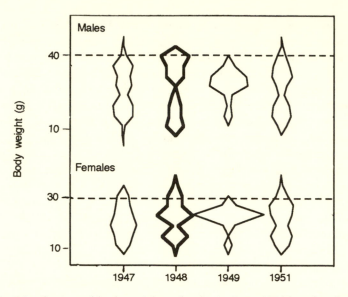

Figure 5.3. Distribution of body weights of voles in June. Widths of each figure are proportional to the percentage in each 4-g body weight class. Thick outline shows peak year. Note the absence of heavy adults during the decline in 1949. Weights of females corrected for pregnancies. From Chitty and Chitty (1962b: Table 2).

Figure 5.4. Maximum body weights of Townsend's voles at any time during each 12-month period of a prolonged low phase (1971–1974) and subsequent increase and decline. Pregnant females included. This decline followed in 1976 among heavy animals without an intervening peak year. N = numbers in spring. From Krebs (1979) and Chitty (1987).

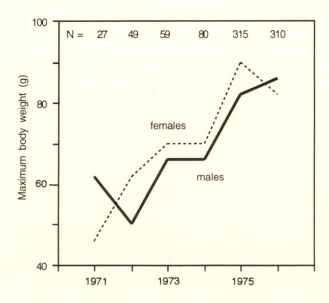

expected from the absence of overlap between age classes already seen in Figure 2.5. Female body weights were distributed in a similar way. To make up for our own lack of data on the low phase, I present Figure 5.4, which shows that no heavy voles were present throughout an unusually long period of low numbers in spring 1971–1974 near Vancouver.[14] On this occasion heavy animals were present during the decline, which followed recovery without an intervening peak year. Heavy animals were also present during a 2-year decline among lemmings (Figure 9.8).[15]

Body weights in other months at Lake Vyrnwy were less clearly related to the phase of the cycle—in September, for example, they were similar whether the population was going to go up or down next year, and except on one occasion we found no other way of predicting what would happen 6 months later. The exception was in 1954–1955, when an increasing population went on breeding at least until November. It reached a peak the next spring, had already weaned some young by March 15, 1955, and may have bred all winter. Also, the overwintered animals were about 20% heavier than those in a population that was declining at the same time. Furthermore, although some of the latter females had become pregnant before the middle of April 1955, no young entered the traps until June, a pattern similar to that in 1947 (Figure 5.3)

The gist of these observations is that adults in the increase phase are sometimes as heavy as in the peak year, differ in breeding both earlier and later than usual, and almost always leave offspring that survive well, grow well, and become unusually heavy in the peak spring. By contrast, animals born during a high peak survive poorly, seldom or never reach maximum size in the field (in spite of the large size of their parents),[16] and leave descendants that may remain low in body weight for a year or more of low numbers. We can perhaps be excused for thinking this low body weight condition was pathological. But whatever its explanation, this condition makes it impossible to use body weight as a criterion of age. Nor can low body weight be interpreted as due to poor food, as it was as late as 1977, when Myllymäki wrote: "The few animals that survive the winter show low growth rates in early spring, and the start of reproduction is retarded. This points to undernourishment during the wintering period. . . ."[17] Ideas travel slowly in population ecology.

5.3 Prewar Body Weights and Reproduction

To see what is general in what is particular and what is permanent in what is transitory is the aim of scientific thought.

Whitehead[18]

Like all observations, whether in physics or evolutionary biology, our observations on body weight were unique; so, to see if events at Lake Vyrnwy

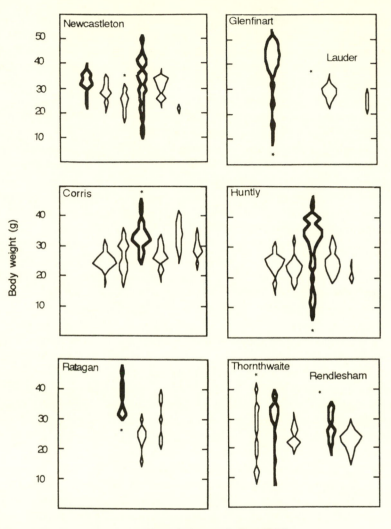

Figure 5.5. Proportion of male voles in each 4-g weight class in April 1929–1935 during the cycles shown in Figure 2.2. Thick outlines show peak years; * shows 1932.

were more than a local peculiarity, that is, belonged to a *class* of events, Helen and I had a look at Baker and Ranson's data for 1930–1931 and at those for later years. In all but two populations, peak years were associated with the presence of larger animals than in the years immediately before or after (Figure 5.5). The exceptions were (1) Glenfinart, for which there were no data for other years but whose voles included some of the heaviest observed, and (2) Thornthwaite, where the distributions were alike for increase and peak years.

Body weights in some peak years wcre not as high as in others: the relatively low peak at Newcastleton in 1930 was associated with relatively low body weights, and the peak body weights at Rendlesham were also fairly low. At Lauder, the only data are for the first and third years after a peak, when body weights were low. At Corris, numbers were only slightly higher in 1932 than in other years, but body weights were appreciably higher than in the 2 years before and in 2 of the 3 years after.

Having young animals in the spring sample, as happens in some peak years, lowers the mean body weight of the population. So unless adults can be distinguished from young, we must look at distributions rather than means. Later on, however, Helen was usually able to distinguish young from old animals of the same body weight.[19]

We had now discovered that qualitative changes were no mere peculiarity of voles at Lake Vyrnwy and could try to clear up some of the puzzles left over from the early work. A hint that the length of the breeding season might be affected by population processes had come from its early end during the peak year 1937 at Lake Vyrnwy and more convincingly from the late ending and early start during the increase of 1954–1955. We now thought we could explain why, in spite of shorter daylength, breeding had ended later and begun earlier in the north than the south. Baker and Ranson had been surprised to find the most northerly population, at Huntly, was still breeding in December 1931; as a result, numbers went up that winter instead of down (see Figure 2.2). And at Newcastleton in 1930, where the population increased to a peak that autumn, breeding had started unusually early, 8 of 12 females having been pregnant in March, none elsewhere. We therefore assumed that the winter and early spring breeding was due to the quality of the animals present during the increase phase. Studying the physical factors while neglecting the state of the cycle, though understandable in the early 1930s, was a fault to be avoided in the future.

If Baker and Ranson had suspected that body weights varied with the cycle, they would not have pooled their data for three areas; but at this early stage, methods of monitoring the populations were not known to be trustworthy, were sometimes applied only once a year, had not yet shown there were such things as cycles, or gave results that made no sense at all. But Baker and Ranson were also unlucky in ending their study before the 1932 peak at Huntly and missing the violent fluctuation at Glenfinart. They were also unlucky that the peak at Newcastleton in 1930 was not pronounced, nor was the high year at Corris.

Now that we knew that voles might breed into the winter before a peak, we could place more confidence in the apparent anomalies of April populations being higher than those in the previous September. At Huntly, the catch went up from 38 in September 1931 to 86 in April 1932; at Glenfinart, voles were apparently more abundant in spring 1935 than in the previous autumn, and the forester told us that voles had indeed bred that winter. I remember Charles being sceptical about this report, which was contrary to Baker and Ranson's findings, based on such an impressive number of post mortems. Their samples, however, had been restricted to 2 years of a 4-year phenome-

non, and like the inference that young animals remain immature in their first
year, their conclusion illustrates the dangers of going too fast in population
research. Even in 1942 the relation between breeding and phase of the cycle
was still unknown. Winter breeding at Glenfinart in 1933–1935 seemed to
be an exception, probably related to the climate in that part of Scotland
rather than to the phase of the cycle;[20] which is not to deny that, besides a
phase of increase, suitable weather as well as other conditions may be necessary for winter breeding. The most dramatic evidence of the phenomenon
was yet to come.

With further data, the association between peak years and high body
weights had become clear; but we discovered it only because we knew from
later work what to look for. We have already discussed the role of prediction
in testing ideas. Retrodiction fulfills the same role, though sceptics will feel
happier when confirming evidence is obtained after publication than before.
This will remove any possibility that the observations gave rise to the idea
but were dressed up to look like predictions. Indeed, unless published ahead
of the data and stated "in terms of *determinate* empirical operations,"[21] predictions do not fulfill their logical function of guarding against the fallacy
of selection.

Since 1937–1938, variation of body weight with phase has turned up
again and again, not only in other species of voles but in water voles,[22] lemmings,[23] and snowshoe hares.[24] The association between high numbers and
high body weights is especially well documented for the latter (Figure 5.6),
but with no discussion of the anomaly that body weights were highest when
there was least food and lowest when there was plenty. Predation, which was
inferred to account for later stages of the decline, does not explain the low
body weights,[25] nor did the similarities between vole and hare cycles persuade

Figure 5.6. Mean body weight of adult and young snowshoe hares in November–
January during the cycles in numbers shown in Figure 1.9. From Keith and Windberg
(1978: Figure 7).

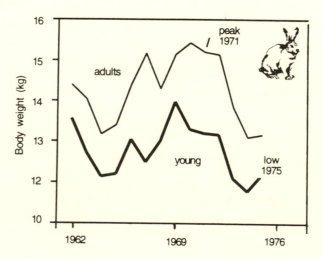

the authors to consider an explanation in terms of social behavior. The correlation between high density and heavy damage to the vegetation seems to have been more persuasive.

Because no two events are identical, it is often difficult to decide whether to split them because of their differences or lump them because of their similarities. The frequent occurrence of these high body weights adds to the probability of there being some process common to all cycles. Lumping therefore increases the opportunities of finding evidence against a point of view, though one's critics are more likely to condemn such generalizing as unwarranted speculation.

5.4 A Two-Year Decline

Yet once again we must wonder both in the past and in the present that the human mind, which goes on collecting facts, is so inelastic, so slow to change its framework of reference.

Butterfield[26]

Thomas Park, University of Chicago, made an indirect contribution to the vole work during his stay at the Bureau in 1948, when he made me promise to write a review of mark-recapture methods. Being under no illusions about my mathematical skills, I asked George Leslie to help me; but instead of reviewing existing methods, he invented new ones, which to Tom's disgust, gave me an excuse to renege on my promise to write the review. Denys Kempson built us a machine similar to that used in Bingo, and for my skill at turning the handle I became junior author of the first of three papers, which included mark-recapture data obtained by using numbered counters.[27] As a tribute to George's genius, this paper was given pride of place in the jubilee issue of Biometrika, and for supplying the field data below, Helen and I became joint authors of the third paper, on the application of the methods.[28] In science, as in other creative fields, there's no telling what spark may kindle a new idea. The secret of success is to create an environment in which people knock these sparks off one another. Among many distinguished pilgrims to Oxford's ecological Mecca, Tom Park was one from whom we reaped this sort of benefit.

One of the next things to do after the war was to replicate the one and only mark-recapture study on which so many conclusions were based. Helen and I chose for our study area a strip of rough grass, about 410 yards long and 40 yards wide, between Lake Vyrnwy and the road around it. The area was bounded on three sides by unsuitable habitat into which voles could disperse but from which few if any were likely to immigrate. We began the study in May 1948, a peak year, and followed the decline through a low point in July 1949, and recovery to fairly high numbers that autumn. We then called off our field trips until spring 1950, when we expected further increase to be on its way.

The population changes differed from those in the early study in three ways: two minor and one major. The first minor difference was in the peak year, when the early-born females and at least some of their brothers survived and matured. This difference was probably because the initial breeding population in 1948 was less dense than that in the peak year 1937. Here, then, was further evidence that Baker and Ranson were wrong in believing that young voles fail to mature in their first year. The second minor difference was that in spring 1950 both sexes disappeared at the same time, although in 1949, as in 1938, the males declined some weeks before the females.

The major difference between this study and that of 1936–1939 gave me a nasty shock. I thought the scarcity of July 1949 marked the end of the cycle and that good recruitment that autumn marked the start of the phase of increase. Because of other commitments we left the voles alone until April 1950, when, to our embarrassment, we found most of them gone. We had missed the chance to study another, apparently different, kind of decline. From a moderate peak in 1948 this population had declined only part way in 1949, and in spite of recovery that autumn, had gone the rest of the way in 1950. To illustrate this pattern of decline, I turn to a well-documented Swedish study[29] of a supposedly noncyclic population, in which the partial recovery of 1984 (Figure 5.7) resembles the recovery of 1949 at Lake Vyrnwy, and the scarcity of May 1985 resembles that of April 1950. The main difference between the two instances is that scarcity was greater and lasted longer in ours than in the Swedish study, as recovery at Lake Vyrnwy was delayed an extra year. Low numbers in spring in spite of recruitment the

Figure 5.7. A two-year, Type-H decline among voles in Sweden. From Erlinge et al. (1990: Figure 4).

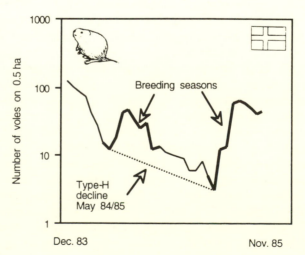

previous autumn, must be regarded as one of the criteria for self-regulation, whether or not we decide to call a population cyclic. But basing this decision on autumn instead of spring numbers can easily obscure the issue. It's the contrast between spring numbers that's typical of cycles. If ignored, this criterion can lead authors, including my own colleagues,[30] to claim their populations are noncyclic (see Figure 12.7).

In hindsight it's easy to claim that I should have been prepared for the delayed disappearance of this population. However, it's also easy to forget how little we knew at the time and how long we were taking to pick up the traces dropped during the war. Several of the "facts" that are popularly supposed to speak for themselves had remained silent. First, this two-stage type of decline had been reported by Hamilton,[31] though neither he nor Charles[32] had commented on this peculiarity, nor had I in the thesis I had just submitted.[33] Second, the Newcastleton population had also behaved this way; but the published account had stopped short at the temporary and supposedly slight autumn recovery;[34] meanwhile data for the following year (which I myself had helped collect!) were lying unanalyzed in our files. Third, further unanalyzed data might have shown me that the Huntly population (see Figure 2.2) had also taken 2 years to reach rock bottom (and had stayed there another year in spite of recovery the previous autumn). Fourth, I already knew that from one and the same autumn density a population might or might not decline (see Figure 3.2); so I should have known that population density alone was useless for predicting population trends. Fifth, and finally, Green and Evans had described a 4- or 5-year decline in snowshoe hares, which should have alerted me to the possibility of a similar sustained decline in voles. These now look like obvious clues; but at the time, they had not been interpreted—a clear example of the importance not merely of reporting one's findings, but of trying to explain them. It's easy to overlook significant evidence until one has learned the hard way what to look out for and what preconceptions to beware of.[35] I console myself with this thought when I see people ignoring crucial evidence I so vividly remember collecting. One such vivid memory is the collapse of this population in spite of recovery the previous year.

Much of this work was handicapped by the difficulty of working half a day's journey from Oxford. With our low-key budget, it was hard to justify travel and hotel expenses when animals were so scarce that the only reward for 5 days work was catching half a dozen voles and putting a single spot on a graph. So knowledge about the low phase was still nonexistent. Yet we could not work closer to home because we imagined (wrongly) that to solve the problem we needed large samples of animals for post mortems. We also thought it unlikely (wrong again) that voles had cycles except in areas in Scotland and Wales that supported high numbers. So visits were neither as frequent nor as regular as they should have been, which complicated the new methods of analyzing mark-recapture data that George invented to go along with this study.

5.5 A Botanical Survey

A prejudice is more easily detected in the primitive, ingenuous
form in which it first arises than as the sophisticated, ossified dogma
it is apt to become later. Science does appear to be baffled by
ingrained habits of thought, some of which seem to be very difficult
to find out. . . .

Schrödinger[36]

At the time of the last study we were still hard up for ideas, and as a long
shot thought the voles might somehow "condition" their food in the way
that flour beetles make their environment toxic.[37] Accordingly, we sought the
help of two botanists from Aberystwyth: Mr. W.E.J. Milton of the Welsh
Plant Breeding Station and Mr. J. Lewis of the Department of Animal Health.
Mr Milton thought it conceivable that a great concentration of droppings and
urine might give rise to a highly nitrogenous herbage containing an excess of
amino acids, and that these might make the females sterile or have other
adverse effects. We took this idea seriously for a while but refrained from
embarking on a biochemical study that would have involved long-term work
outside our competence and justified by no more than a wild guess. This was
a judgment of relevance based on the difficulty of seeing how toxicity could
account for the demographic peculiarities of a decline or could act uniformly
across the diversity of plant communities from which voles disappeared at
the same time.

Instead, we asked Mr. Milton and Mr. Lewis to look for evidence of food
shortage. Their studies covered the peak year 1948 and the declines of 1949
and 1950. The upshot of their long and detailed measurements was that the
vegetation at Lake Vyrnwy was in much the same state as that at similar
places where voles were scarce or absent. More specifically, they concluded
in June 1950 that "There has obviously been no shortage of edible green
material on the wet soil areas, and it is also unlikely that the dry areas have
been devastated in the sense of a real food shortage."

Their conclusions were helped by the unusual diversity of the vegetation
within our study area: scanty in some places, luxuriant in others, and green
all winter along the grass verges of the road and beside streams flowing into
the lake. By the end of winter as little as 3% green vegetation was left in
some places, compared with 35–85% in others, and as in the prewar work,
voles disappeared at the same time regardless of the amount or species of
herbage in which they were living. We therefore decided that, as explanations
for the decline, neither quantity nor quality of food was worth further study.
(Both, of course, are relevant to explaining why voles differ in abundance
between good and bad habitats and good and bad years.)

Mr. Milton and Mr. Lewis unfortunately considered their findings too un-
remarkable to be worth publishing. Their manuscript is deposited in the Elton Library in the Department of Zoology, Oxford University.

5.6 Survival in the Lab

Negative facts when considered alone, never teach us anything.
Bernard[38]

A fourth prewar observation that had to be replicated concerned survival in the lab. We had given up expecting animals from a declining population to die of disease in the lab in parallel with those in the field. However, according to my new hypothesis, animals born in peak population were less viable than those born during the increase. We needed to compare the survival of these two classes of animals. So, in 2 successive springs—1948 and 1949—we collected about 50 animals from Lake Vyrnwy and kept them, one to a cage, until all, or most, had died (Figure 5.8). The overwintered animals in the peak spring of 1948 had been born in the increase phase and were supposed to be more viable than their offspring born during the peak and removed to the lab during the decline of spring 1949.

The first surprise was that in both years these wild-born animals, already several months old,[39] survived longer than expected from the Leslie-Ranson results with domesticated stock. The second surprise (and disappointment) was that survival was much the same in both years. Looked at in retrospect, these results were the first nails in the coffin of the maternal stress hypothesis; but at the time they were not sufficiently discouraging to make me switch to another track, even if I'd known where to find one. A reason for not being discouraged was that failure could be blamed on faulty experimental design.

Figure 5.8. Survival in the lab of male voles born during increase in 1947 (and captured during the peak in March 1948). Contrary to expectation, results were similar for voles born during the peak (and captured during the decline in March 1949). In the field, however, survival was worse in 1949 than 1948.

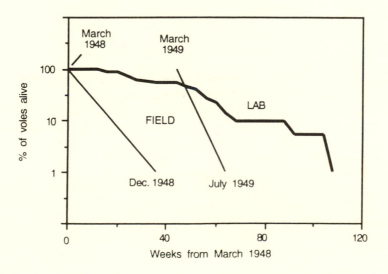

Ideally, the two cohorts should have been started in the lab at the same time so that they were living under identical conditions. However carefully one tries to standardize the environment, differences are likely to creep in year by year. Technicians change, accidents happen (such as lights or heating going off), methods of husbandry get better or worse in unsuspected ways, and in the rather primitive animal room, which is all we had, the environment varied from season to season and year to year. Second, the decline of 1949 was so much less severe than that of 1938 that the postulated effects of reduced viability had perhaps been too subtle to detect. This sort of rationalization may or may not turn out to be justified; *ad hoc* arguments have their uses in keeping up morale until something better comes along.

To avoid the error of comparing results in 1 year with those in another, I would have had to find neighboring populations living in comparable habitats but in different phases of their cycle. On this occasion I was out of luck. And when such populations turned up a few years later, the results, though in the expected direction, were equally inconclusive because the declining population was heavily infected with vole tuberculosis.

A final reason for believing these results were merely negative rather than contradictory was the realization, implicit in the hypothesis but not explicitly stated until later, that I had removed these animals from their normal mortality factors. A necessary condition was, therefore, missing. I should have provided some kind of stress instead of making conditions as favorable as possible. Bringing animals into the lab is equivalent to providing hospital care for the elderly and handicapped and being surprised at how much longer they live than those in the outside world. In other words, a change in age or viability is not a sufficient explanation for a reduced survival rate. Innate and environmental factors must both be considered, as Medawar makes clear when he defines

> Senescence . . . as that change in the bodily faculties and sensibilities and energies which accompanies ageing, and which renders the individual progressively more likely to die from accidental causes of random incidence. Strictly speaking, the word "accidental" is redundant, for all deaths are in some degree accidental. No death is wholly "natural;" no one dies *merely* of the burden of the years.[40]

To apply this definition to my own work, I had only to substitute 'change in quality' for 'senescence' and "hostility" for 'ageing.'

After the failure of these two life-table experiments, I did little further work on survival, except as a by-product of other work; it took too long and tied up too many cages. However, it's an experiment that should be tried again, but designed to show whether breeding success in captivity varies in parallel with that in the wild. A few later observations suggested that it might; but it was many years before the idea was corroborated.[41]

The results of the first year's observations are shown in Figure 5.8. In addition, I introduced 18 lab-born voles into the system to see if my husbandry was up to scratch. Indeed it was, for at 71 weeks, the expectation of life was more than twice that of Richard's prewar stock.[42] I believe the improvement was due to my giving cabbage twice a week and cutting out vitaminized oil. I had also stopped cleaning the cages as often as Richard used to. At the time, this change in procedure obviously had no ill effects on the voles; later on, however, it caused pain and anguish not only to the voles but to me.

5.7 Types of Decline

Real science exists, then, only from the moment when a phenomenon is accurately defined as to its nature and rigorously determined in relation to its material conditions, that is, when its law is known. Before that, we have only groping and empiricism.

Bernard[43]

We had now studied two slightly different types of decline, which we assumed were variations on a common theme. But some populations presumably declined for a variety of reasons unrelated to cycles. How was one to distinguish between cyclic declines and those due merely to lack of breeding in winter or to accidents of various kinds? How was one to avoid getting bogged down in the impossible task of finding a common antecedent for declines that had nothing in common but a nondiagnostic drop in numbers?[44] Second, how was one to distinguish between factors that merely modified a cycle and those that were essential to it? According to the prevailing wisdom, if disease was the cause[45] (C) of the cyclic decline (E for effect), then symbolically C→E. I now proposed that C and E were both effects of an antecedent condition X, and that C merely modified E, thus

$$X \nearrow_{\searrow} \begin{matrix} C \\ \downarrow \\ E \end{matrix}$$

The need to define appropriate test instances is common to all branches of science; critical tests are otherwise impossible. No hypothesis predicts events outside its own universe of discourse. Defining (or essential) characteristics are those without which the thing, event, or phenomenon would be unrecognizable; accidental (or accompanying) characteristics are those that can be omitted from the description.[46] A triangle is defined as a figure with three sides; its infinite varieties of size and shape are accidental characteristics. We need to recognize characteristics among cycles that are analogous to the essential three-sidedness of a triangle. We can start by describing types of decline.

Among variants we recognized three recurrent patterns.[47] Doug Middleton[48] had found a population that completed its decline at some unknown time during the winter: Type M, which, in the absence of details, can be explained in terms of anyone's preconceptions; Gillian Godfrey[49] had found

Figure 5.9. Helen giving one of our two papers at the original International Theriological Congress in Brno, Czechoslavakia in 1960. She presented the body weight data for 1945–1960; I presented prototypes of Figures 5.10 and 8.5.

one that declined during the breeding season without recovering in autumn: Type G (Figure 5.12) and Hamilton[50] had found one that started the same way but recovered temporarily before autumn and reached rock bottom the next year: Type H. This pattern was confirmed to my embarrassment in 1949–1950. The Newcastleton decline of 1934–1935 was also Type H, though we did not recognize it at the time, and more recent examples are shown in Figures 5.7 and 12.7. The unexplained feature of this pattern is the persistence of low spring numbers for 2 or more years—in sharp contrast to later increase from equally low numbers. Other types included the pattern at Lake Vyrnwy (Type L), where the decline began in late winter and accelerated in spring, that at Corris, which was largely seasonal, and those at Glenfinart and later on that in Wytham that skipped a typical low phase altogether. These were some of the numerical patterns we now had to incorporate into a fuller, biological description of a cycle.

5.8 Defining a Cycle

Nothing ever really recurs in exact detail. No two days are identical, no two winters. What has gone, has gone for ever. Accordingly the practical philosophy of mankind has been to expect the broad recurrences, and to accept the details as emanating from the inscrutable womb of things beyond the ken of rationality.

Whitehead[51]

As long as cycles are thought of as purely numerical changes from one year to the next, almost any explanation will fit. However, the changes in which we are interested are no mere numerical events but biological processes that typically recur in an interrelated sequence. The migration of the lemmings, the regularity of the cycle, the association with disease, and the synchrony of independent populations were some of the characteristics first noted. But it soon turned out that these properties, though typical, were only sometimes associated with recurrent cycles, and being unknown for new areas, could not be used to distinguish them from miscellaneous changes in numbers. Aspects that had once seemed to define the phenomenon had become irrelevant or unreliable as knowledge grew.

By 1950 we knew many of the characteristics of a cycle in numbers; but variants kept cropping up, some of which were not published until 12 years later,[52] and others followed in 1971.[53] Some of these variants were due to differences at the peak between high and moderate breeding densities, some due to differences in length of the low phase. Using regularity as a criterion of cyclicity was becoming less and less useful; nor is a cyclicity index desirable;[54] instead, using a sequence of biological characteristics seemed a better way of recognizing a cycle.

Figure 5.10 summarizes some of the characteristics of the phases of a 4-year cycle; but cycles are more difficult to recognize when the sequence is shorter, has no refractory low phase, or when three phases, increase, peak, and decline, all occur in the same year. It is also difficult at the time to know whether increases and decreases are cyclic or merely seasonal or due to associated changes in food supply, weather, or other uncontrolled variables. The low phase is also difficult to analyze. From the same low numbers, populations may or may not increase to a peak; so from density alone one cannot predict what is going to happen. Nor do we know how else to distinguish between populations at the end of a decline and those at the beginning of

Figure 5.10. Typical patterns among vole cycles, including four types of decline: G, L, H, and M (See Section 5.7). See also Krebs and Myers (1974).

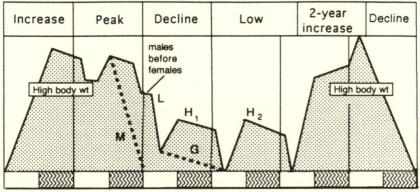

summer   winter

Populations at the end of a decline include two extreme phenotypes, those that grow neither in the field nor in the lab, and those that don't grow in the field but do so in the lab. Perhaps the proportions of these two phenotypes determine how long the low phase will last.

High body weights are a help in enabling one to recognize a peak, but by themselves are unreliable criteria for cyclic behavior.[55] In 1932, when numbers were slightly higher at Corris than in other years, body weights were appreciably higher. Yet by numerical criteria, this population was noncyclic. To add to one's difficulties, body weight is affected not only by variables such as age, food, and reproductive condition, but also by the social environment. Variation in the length of the breeding season is another criterion that may help identify a cycle in voles, also in lemmings (which breed in winter during the increase phase)[56] and in snowshoe hares,[57] which have short breeding seasons at high densities (Figure 5.11). But the most nearly diagnostic features of a cycle are probably (1) a phase of rapid increase from scarcity followed within 2 years by (2) poor juvenile survival; (3) a reduced rate of increase; (4) a decrease that's too sudden and discriminatory to be explained by disease, food shortage, senescence, or increased predation; and (5) by a refractory low phase. Such changes would be fully diagnostic if the sexes disappeared at different times and had lower body weights than at the peak; but diagnosis is less certain if some of these symptoms are lacking or if authors fail to describe them.

For an example of the difficulty of deciding whether a population is cyclic or noncyclic, I turn to another well-documented example from Sweden.[58] As shown in Figure 5.12, data on population size were limited to April and August, and none were given on differences in survival of males and females or young and old. For the following reasons, however, I would regard this population as having had a 5-year cycle: (a) numbers were unusually high in November 1983, (b) the population had a Type-G decline in 1984 (decline throughout the breeding season), (c) spring numbers remained low during

Figure 5.11. Cycles in numbers of snowshoe hares and length of breeding season. From Keith and Windberg (1978: Table 23) and Cary and Keith (1979: Table 8).

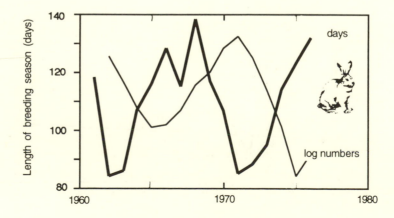

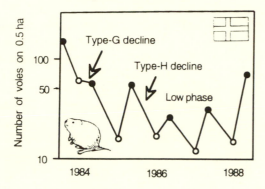

Figure 5.12. Number of voles per half-hectare in southern Sweden in November 1983 (●) and April (○) and August (●) in other years. From Agrell et al. (1992: Table 1).

1985–1988 in spite of recruitment the previous autumns, and (d) young of the year were much heavier during increase in 1988 than in 1987. One difficulty is that body weights of overwintered animals were high not only in 1988 but in 1985 as well, which is unusual but not unknown during a decline. The authors point out that high body weights were correlated with high plant productivity and regard the numerical changes as noncyclic, though admitting the population "has the potential for cyclic fluctuations." We should not get bogged down quarreling over labels, however, for the real bone of contention is over the role of social behavior in causing population changes, whatever we call them. In denying its role, the authors observe that there's "no relationship between density and body weight." Their observation is correct, their inference is not. Indeed, the lack of relationship between density and body weight is characteristic of cyles and is one reason for believing that recurrent differences in body weight are due to recurrent differences in behavior. Whether populations are or are not cyclic is irrelevant to deciding the role of behavior; the two aspects cannot be dismissed in the same breath.

One other difference in emphasis must be mentioned. The authors focus on population growth during the breeding season instead of from one breeding season to the next. This ignores one of the puzzles about these fluctuations, whether cyclic or noncyclic,[59] namely, that recruitment in 1 year so often fails to increase the next year's breeding population. At present we must agree to disagree on matters of interpretation, while agreeing that more is going on biologically than is revealed by mere numerical changes. The problem is to know which of the biological changes are mere symptoms and which are fundamental characteristics. Changes in body weight and length of the breeding season, for example, may be largely effects of the social environment, but may also be due to natural selection, and both may be confounded with effects of other variables. Until we can distinguish between symptoms and causal factors we shall remain in the state of mind that Bernard calls 'only groping and empiricism.' We've been too long at this stage.

Behavior, Physiology, and Natural Selection, 1949–1961

And generally let every student of nature take this as a
rule: that whatever his mind seizes and dwells upon with
peculiar satisfaction is to be held in suspicion.

Bacon[1]

6.1 Selye, Christian, and I

By 1950 we had replicated the prewar observations and made as much use
as we could of purely descriptive evidence. To test the implications of the
maternal stress hypothesis we now had to do manipulative experiments. The
first part of the idea—that voles have a form of behavior that keeps their
numbers in check—could not be corroborated by direct observation in the
field. If they had been birds fighting for territories, the task would have been
easier. But as voles are seldom seen in nature, the only way to check a behavioral hypothesis was indirectly, which meant adding extra links to the chain
of reasoning. As some links were weak, it was prudent to think up a variety
of implications. For example, one could study the behavior of voles in captivity and be wrong in thinking they behaved that way in nature. Or one could
manipulate wild populations and disrupt their social structure in more ways
than intended. Some results might confirm, others refute one's preconceptions; but everything would be grist to the mill and might some day point in
the right direction.

The second part of the idea—that the offspring of stressed females were
abnormal—was even harder to test. It could not be dismissed as easily as the
other explanations for cyclic behavior. We needed some way of measuring
this condition in the field, and the best approach seemed to be to measure
liver glycogen in the hope that voles, like snowshoe hares, suffered from
shock disease. We also needed to produce the condition in the lab, find out
if it resembled that in nature, and if possible breed from low-phase animals.
It was hard to know how otherwise to proceed. Then came the light on our
road to Damascus.

Until 1949 I had not heard of Selye's ideas about stress. Although his
theory of the General Adaptation Syndrome had been published in a brief

note in 1936, his ideas did not become widely known until later.[2] I first heard about Selye at a meeting of the Society for Experimental Biology, when Dr. M.R.A. Chance of the University of Birmingham suggested his ideas might provide the theoretical framework for my otherwise bald and unconvincing narrative. Shortly thereafter Dr. Wes Whitten made the same suggestion to John Clarke, and later still, John and I attended a lecture given by Selye himself in the Radcliffe Infirmary. The gist of his ideas is that organisms respond to stress of all kinds with a nonspecific reaction (the G.A.S.), but that under prolonged stress an animal becomes exhausted and susceptible to a variety of further insults.

Then in 1950 came an important article by John Christian.[3] He had beaten us to the punch in breaking with current preconceptions. His article was a great stimulus to new ways of thinking, but failed to address the difficult part of the snowshoe hare story, namely, that death rates, instead of getting better as numbers declined, got even worse. The animals with the highest death rates—at the bottom of the cycle—had not been overcrowded. Christian was wrong in overlooking this difficulty; I was wrong, in the way I got around it;[4] but we both made a break—albeit a small one—in the prevailing conceptual logjam.

Thus, at the beginning of the 1950s we were no longer hard up for ideas and at the time had no doubt we were about to solve the cycle problem. Four independent ideas had converged: George Leslie's idea, based on laboratory data, that declining populations were older than usual; my idea, based on field work, that voles were congenitally abnormal; the Minnesota workers' idea, based on trapped and captive snowshoe hares, that they suffered from shock disease; and Selye's ideas about stress, based on clinical evidence and applied by Christian. In fact, all four can be looked upon as variations on the single theme that increase in death rates is due to adverse changes, not only, or even necessarily, in the physical environment, but in the animals themselves.

The belief that we were about to solve the problem made this an exciting time, and seeing if there was any truth in the maternal stress hypothesis kept us busy for the next 7 years. "The advantage of this hypothesis was that it was readily testable; the disadvantage was that year after year it kept giving encouraging leads, all but one of which . . . turned out to be will-o'-the-wisps."[5] This did not mean we were making no progress; on the contrary, without the hard work of the next few years we would still be in the intellectual time warp of the 1930s. The trouble was a classical one: that crucial experiments seem crucial in retrospect only.[6]

At this stage, while Helen and I continued our field and lab work, I was lucky enough to become the supervisor of two D.Phil. students, Gillian Godfrey and John Clarke, who studied aspects of the problem that would otherwise have been neglected. Meanwhile, time was running out for me to submit my D. Phil. thesis, which I finished with about 5 minutes to spare before the final extended deadline in 1949. Although such tardiness is reprehensible, it made me sympathetic later in my career toward students displaying the same fault. Another 3 years elapsed before the thesis was revised, refereed, and published. Then 2 more years went by before anyone took public exception to its conclusions.

Although it's as well to know what you've done before proceeding along the same possibly wrong track, writing reports and papers conflicts with the need to keep going. Once interrupted, the sequence of natural events is ruined for good—another disadvantage from which the laboratory zoologist suffers less than the ecologist. In my case the contest was one-sided, owing to my greater pleasure in doing things than writing about them. At this stage of my career, however, I'm having fun writing this book.

6.2 Sparse Populations

Scientific investigations and experimental ideas may have their birth in almost involuntary chance observations which present themselves either spontaneously or in an experiment made with a different purpose.

Bernard[7]

Gillian Godfrey, an Oxford graduate, took her D.Phil. under my supervision. She later married Peter Crowcroft, according to whom she had to rewrite those parts of her thesis that supported the views of her supervisor. David Lack, one of the only two examiners called for by the Oxford system, had other objections as well; but rightly or wrongly, some of us felt that Gillian was an innocent victim of the dogs of war. The incident did nothing to sweeten relations between David and the Bureau.

From 1949 onward I was lucky enough to have at least one graduate student working on some aspect of the cycle problem. One of my first students, Gillian Godfrey began studying the early life of the vole in two low-density populations living half a mile apart, on The Dell and Rough Common in Wytham Woods, a short distance from the Bureau.[8] Despite many hours on hands and knees, she found hardly any nests and attracted no tenants to her various designs of nest box.[9] So she solved the problem by being the first to use radioactive tagging methods to find nests,[10] an idea that developed over the years into a highly sophisticated technique (Figure 6.1).[11]

The two populations increased in 1950; one decreased and one increased in 1951; and one decreased and one remained scarce in 1952. As in other populations, the season's young experienced unaccountable losses during the breeding season.[12] One of these declines differed from those so far studied, as numbers fell throughout the breeding season without recovering before the onset of winter. At Newcastleton, by contrast, the summer decline in 1934, and that at Lake Vynrwy in 1949 had been temporarily reversed.

Gillian also found wounds on some of her animals and low weights among two litters when they left the nest. Neither of these observations has been

Figure 6.1. Xavier Lambin searching for vole nests at Ladner, British Columbia, in 1990. See Lambin and Krebs (1991). Photo by Sherry Kendall.

replicated, as far as I know, on other short-tailed field voles. (The low weaning weights were perhaps due to the deaths of the mothers or to their being confined in traps. Young animals that actually enter the traps are normal in body weight.)

We now knew that cycles went on even in low-density populations;[13] and the continual decline throughout the breeding season added a variation I later called type G, after the G in Godfrey.[14] In population ecology we must be content with small honors. As a population study (which was not how it started out), this was one of the most economical. The two areas, being close together were exposed to the same weather and predators, which included foxes, stoats, weasels, barn owls, tawny owls, and kestrels; also to three species of shrews. As the declines occurred in different years and affected young more severely than old, and males more severely than females, special pleading would be required to explain the differences in terms of bad weather or predation. And boldly committing lèse majesté, Gillian stated that "There was little support for the suggestion advocated by Lack (1954) that food shortage can account for mortality of the type described here."

6.3 Direct Effects of Behavior

The result of the experiment is the difference between two sets of readings (or two sets of phenomena or two events) . . . In everyday life, of course, we speak of the causes of events, phenomena, or states of affairs, but the cause we have in mind, when analyzed, usually turns out to be the cause of a difference between what was and what might have been; between what did happen and what might have happened if the antecedents had themselves been different.

Medawar[15]

The characteristic method of science . . . [is] that objective, communicable, and confirmable statements can be made not about single sense impressions but only about pairs of these. . . .

Born[16]

John Clarke, an Australian Rhodes Scholar and Oxford graduate, also started his D.Phil. in 1949. I owe him much for his collaboration at the experimental and philosophical ends of the scientific spectrum. When writing up his thesis, John, being busy with a new teaching job, had no time to be brief, and the massive length of his thesis provoked the only complaint from Peter Medawar, his external examiner. "Clarke," his inquisitor began (running his thumb, as befits a bridge player, up the thesis pages) "Is there some merit in bulk?" John: "I thought I had something important to say," a claim that was confirmed when the oral exam set a new speed record. Niko Tinbergen, the internal examiner, had a kinder, gentler approach to a nervous candidate, few of whom are lucky enough to be examined by two future winners of a Nobel Prize.

Between us, John and I tackled the problem of the direct and indirect effects of hostile behavior. Studying direct effects offered a reasonable chance of short-term success, so was suitable for a D.Phil. thesis. Studying indirect effects, supposing they existed, was to take a risk suitable only for someone with a secure job and no pressure to publish (both being among the advantages of working at the Bureau). So this was my part of the problem.

Before beginning this work we knew that putting strange voles together might result in fighting, often to the death, but the various forms of aggression and defense had not been worked out. With the help of a movie camera, John made sketches of encounters between strange voles (Figure 6.2).[17] He also became the first to carry out vole experiments along classical, comparative lines, using littermates as controls for animals stressed through contact with strangers.[18] His method was to introduce test animals into the cage of a pair of known aggressive voles. After subjecting them to a number of encounters, John killed the animals and described changes in the weight of the adrenals, thymus, and spleen. As expected from Selye's ideas, the adrenals became heavier and the thymus lighter, but an increase in the weight of the spleen was unexpected: it became approximately double that of the controls. John thought this might have been due to pathogens kept in check in normal animals. But whatever the explanation, these findings supported the hypothesis that fighting affected the physiological condition, at least of animals getting the worst of it.

Figure 6.2. Agonistic activities. (a) Subordinate vole before approach of dominant. (b) Approach of dominant (right); squatting retaliation of subordinate. (c) Dominant retires, subordinate licks hands and brushes nose. (d) Dominant returns; each vole bares teeth and squeals. Toilet activity when an aggressive vole is just inches away seems most inappropriate; it is probably a 'displacement' activity arising from conflict between retaliation and flight. From Clarke (1964).

John also set up two confined populations living outdoors in Wytham Woods in large brick enclosures (one called the 'snake pit', Figure 6.3) "under conditions in which it could be reasonably assumed that the only factor limiting their expansion was strife between their members." [19] As the populations soon stopped increasing, the results agreed with expectation. However, numbers leveled out at far greater densities than those in nature, and at first I was more than a little disturbed at finding how high they rose without producing the expected decline. Mere high numbers were obviously not sufficient; but it made sense to suppose that the required degree of crowding varied with habitat and that once this point had come, a decline could not be far behind. We did indeed think we saw one approaching, but this was probably wishful thinking, as we learned many years later; and if John had not had to wind up this part of the work he might have realized that no decline was to be expected. In which case he would have robbed Charley Krebs of the glory of having the 'fence effect' named after him.[20] It might have been named the 'Clarke snake pit effect,' to signify that enclosing a population destroys one or more of the conditions (such as dispersal) necessary for a decline and forces 'third-class' voles into a single, tattered cluster.

John's work was most encouraging to Helen and me and helped us decide how to test our ideas in the field. It also provided John with an experimental animal on which he and his students worked with great success for many years. His interests were in reproductive physiology, however, and it was many years before a discovery in this discipline had a direct bearing on the cycle problem.[21]

Figure 6.3. 'The Snake Pit,' formerly a 10-ft-deep water storage tank. The floor (67 sq m) is covered with bracken and hay. Also shown are brick nest box, Longworth traps, tin used in collecting animals at censuses, and planks supporting a wire mesh roof.

6.4 Maternal and Other Stresses

Science would never progress if we thought ourselves justified in renouncing scientific methods because they were imperfect; in this case, the one thing to do is to perfect the methods.

Bernard[22]

During my part of the project I, too, made observations on the direct effects of behavior. To discover if offspring were affected by maternal stress I had to find a way of stressing pregnant females and began by using the technique John Clarke had used. I first mated littermate females to the same male, and after intervals of 1–11 days introduced one of them for short periods into the cage of a strange pair. Females that were stressed when less than 1 week pregnant almost always resorbed their embryos: of 31 stressed females, only 4 produced litters, compared with 18 produced by their 31 unstressed littermates. Some introduced in their second week also resorbed their embryos.

I next reduced the severity of the stress so that pairs of voles were left in their own undisturbed cages except for a couple of hours three times per week, when they were connected by a tunnel to two neighboring pairs. Being in their home cages, the animals were better able to defend themselves.

Things now went the "wrong" way; mild interactions were apparently beneficial, as experimental females produced larger litters during and (even after) treatment.[23] In a second experiment, two out of three females responded by producing significantly larger litters than their littermate controls; but one produced litters of the same size, and later replications were unsuccessful or went in the direction originally expected of stressed voles. This unpredictability was presumably due to the difficulty of standardizing the severity of the interactions. So we were stuck with the dilemma that with violent interactions females lost their embryos, and with mild interactions they were either stimulated[24] or unaffected. One does not normally publish irreproducible results; but I've resuscitated this information because there may be something to it. I believe this because of a similar contrast between my later failure to breed voles and deer mice in small cages and the spectacular success of breeding them in colonies (see Section 12.4).

We made three other types of observation on the direct effects of stress. First, we needed to know whether liver glycogen was affected by fighting. This work was time consuming, largely owing to my incompetence as a biochemist, partly to the difficulty of getting repeatable results. Such as they were, however, they made it seem worthwhile looking for shock disease in the field. In one sense this was a waste of time—glycogen levels were normal—but in another sense this discovery marked a big step forward (see Section 8.9).

A second procedure was designed to solve the problem of knowing whether unwounded females were stressed, as some stayed home hidden out of harm's way under cover (which may explain why some females failed to respond to treatment). Disruption of the estrous cycle might, we thought, indicate a disruption of a female's physiology. Unfortunately, the estrous cycle of the vole turned out to be more difficult to follow than it is in rats and mice,[25] and vaginal smears turned out to be useless as a measure of stress. An incidental discovery was that voles could be either spontaneous or induced ovulators—or so it seemed at the time. But some years later, one of John Clarke's students found that changes in the proportion of cells in the vagina of unmated females did not imply a corresponding cycle of ovulation, which he found was induced only by mating.[26] Helen had indeed been unable to relate these changes to the presence of shed ova, but assumed her technique was faulty.

In a third type of experiment we did a double-blind experiment in collaboration with an orthopaedic surgeon to see if stressed voles had symptoms of slipped disks.[27] Dr. Scott had been impressed by the frequency of slipped disks among patients who had experienced mental but not physical stress. We confirmed his suspicion experimentally by showing that in stressed voles the nucleus pulposus of the intervertebral disk became enlarged as it did among his patients. And years later Dr Scott's view about the role of mental stress in causing back injury was confirmed in a massive enquiry among employees of the Boeing Company in Seattle.[28] Our papers were not among the references.

This last experiment was an interesting side issue from the attempt to test the maternal stress hypothesis, which had otherwise made to a poor start; but it was not yet dead in the water.

6.5 Through a Glass, Darkly

The Harvard Law of Animal behavior states that under
carefully controlled conditions animals do as they damn well please.
Anonymous

It was now clear that to elicit the right form of stress we needed something less crude than fighting and wounding among strangers. Indeed, none of the thousands of live and dead field animals handled by Helen and me were seriously wounded. I therefore tried to set up colonies in which home range behavior in the field could be mimicked in the limited space of the laboratory. Somehow, while reducing the home range from 100 sq yd or more to 1–2 sq ft, we had to elicit a form of behavior we had no way of observing in the field. Our first population cages consisted of 12 glass-fronted galleries of 2 × 2 in. cross-section and 4 ft long built one above the other up the walls of the lab. These tunnels represented the runways found in nature; their small cross-section allowed subordinates to defend themselves more easily than in undivided space. A later free-standing model shows how a nest box was provided for each pair of runways.[29]

With this system it was impossible to start a colony with more than one pair of adults because of the severe wounding that ensued. So we next raised a colony from a single pair, which produced behavior so subtle that we never saw it. The original female weaned 12 litters in 280 days, but we found no fighting or wounding even when there were 50 in the family (though two large strangers accidentally introduced one evening had been mutilated and killed before morning). But behavior of some sort was going on within the family, as the original female occupied the less crowded penthouse in the upper of six nest boxes, while the rest of the animals, including the original male, were usually confined to the basement. The distribution in the eighth month was as follows:

Nest box no.	1	2	3	4	5	6	Total
No. of animals	5	3	3	2	7	20	40

In the first 9 months seven animals had died and five had been accidentally killed. I then removed the original pair together with all except one male and one female from each of the 12 litters. As there had been 31 males to 19 females,[30] the colony was now less than half size. The remaining males soon became aggressive, and females that had produced no young in their first few months of life conceived after intervals of 12, 19, 32, 34 . . . days from the time numbers had been reduced. In the previous 9 months only 4 daughters had produced a litter,[31] none of which had survived.

At the same time that we killed over half the colony, we killed several litters of comparable age that had been removed from their parents and kept together since weaning. We weighed the testes of the males and found that those from the colony were underdeveloped,[32] though the animals themselves had grown well. We started similar colonies on other occasions and tried to increase hostility by keeping smaller numbers in twice as much space, by isolating pregnant females in the top or bottom nest boxes until they had weaned their young, by fostering nestlings from other females, by removing animals for one or more days and putting them back (an excellent way of producing a graded response), and by other trial-and-error methods. We hoped to find some intermediate condition between mutilation at one unnatural extreme and absence of home range behavior at the other. Finally we arrived at a system in which three to four females and offspring coexisted with a minimum of wounding and also maintained a spatial distribution.

Figure 6.4 shows a section of a system consisting of 10 adjacent cages each supplied with oats, water, cabbage twice a week, and as much hay as could be crammed into each cage. A hole in the roof of the cages enabled the voles to get into the lower of three glass-fronted galleries, one above the other. The middle and upper galleries were accessible through holes in the floors, and all three were divided vertically into 10 units by removable partitions also with holes in them. In the lower galleries the holes were small (1.75-cm diameter) and prevented large animals from running directly from one cage to the next; instead, they had to take a zig-zag course to the top or middle gallery, where the holes were larger (3.17-cm diameter). Small animals, however,

Figure 6.4. Section of galleries, Bureau of Animal Population, 1954. The large holes were normally placed in the top gallery only.

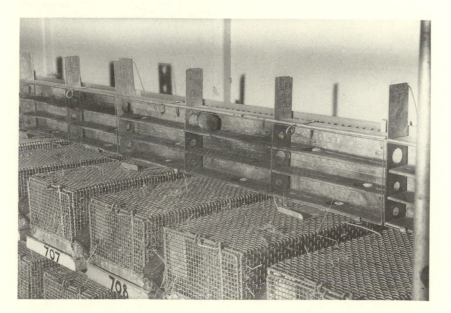

could run the length of all galleries. So by the time a hostile adult had gone upstairs and downstairs the young animal had found sanctuary among the hay in one of the 10 cages. (Many years later I had galleries running round three of the walls of my lab, with a connection across my desk—Figure 6.5.)

When first introduced, pairs were kept apart by solid vertical barriers, after which the females were allowed to roam freely; but to prevent the males from killing one another, I first used medium-sized holes and males that were too large to get through them. Later on I removed all but one male. At intervals throughout the day we plotted distributions in the galleries, the fur of each vole having been clipped or dyed in an individual pattern. Also, whenever we heard a disturbance we rushed over and recorded who was chasing whom and occasionally made a census to discover the whereabouts of animals that spent most of their time hiding in the hay.

With a single male and three or four females this system worked well. Figure 6.6a gives an example in which females A and D remained at opposite ends of the system and females B and C shared the space in between; Figure 6.6b shows the range 2 weeks later, after female A had been found dead and mutilated in cage 4 and most of the young had moved into her space; and Figure 6.6c shows the range after the remaining adults had been removed and the young had spread throughout the system. They had become aware almost at once of the absence of the adults.

By this sort of manipulation we ourselves could produce changes in distribution; at other times we happened to see a fight that by next day had pro-

Figure 6.5. Section of galleries, Animal Care Centre, University of British Columbia, 1986. Photo by Darielle Talarico.

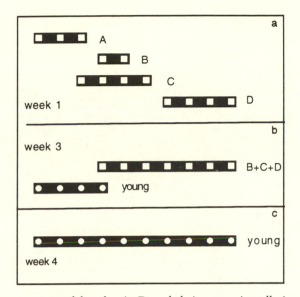

Figure 6.6. Home ranges of females A–D and their young in galleries above cages 1 (on left) to 10 (on right). In week 1, before the death of A, the young remained with their mothers. In week 3 the young moved into A's vacant home range. In week 4, after the removal of B, C, and D, the young took up their own home ranges. Symbols show where the animals were seen most often; occasional sightings (once or twice only) are omitted.

duced a domino effect through an animal's displacing a neighbor, which in turn displaced a neighbor, and so on throughout the system. We also recorded instances in which neighbors that had been hostile one day lived happily ever after and instances in which friends suddenly became enemies. So we inferred that unexplained changes in distribution were due to unseen changes in the relations between individuals.

From records of this sort we concluded that groups might become quite large, with members, not always related, being amicable toward one another but hostile toward neighbors and strangers. We also concluded that behavior might change suddenly, especially if certain animals died, were removed, or were returned after an absence of a day or two. These conclusions prepared us to expect that sudden changes in numbers of field animals might be unrelated to density and unpredictable unless we could measure behavior. It was also obvious that overt aggression was only one of the ways in which voles spaced themselves out, and that other forms of spacing behavior, too subtle to be detected in the lab, would be harder still to detect in nature.

Hostility between groups and compatibility within them suggested that densities might be unevenly distributed outdoors, an implication supported by work in The Netherlands.[33] Figure 6.7 shows a line of demarcation between a large and small clan of voles living in a grassy enclosure. The larger clan evidently refrained or was prevented from raiding the more abundant

Figure 6.7. Grazing by a large and small clan of Continental voles. The sharp boundary line shows that they maintained distinct territories. From van Wijngaarden (1960).

food supplies of its neighbors. Evidence from our lab colonies suggests that a high death rate in nature would prevent this tidy distribution from lasting indefinitely, but that numbers would go on increasing as long as the social order was maintained. Frank, in Germany, also found a line of demarcation between families living in an enclosure.[34] He suggested that vole outbreaks were due to unrestricted increase within such extended families. Preoccupied as we were with the crash, I failed for many years to recognize this contribution to understanding the phase of increase.

Lab evidence had now confirmed that hostility was a plausible mechanism for preventing increase and its absence was a plausible explanation for permitting it. But hostility by itself would set an upper limit to numbers, not reduce them, which was the delayed effect I was trying to produce. So far I was out of luck.

Although perhaps as boring as Crick points out, this description of methods may prove useful to students trying to disentangle the complex relations within captive populations. Some of these relations are relevant to those in nature, others are artifacts of captivity. To avoid wasting time on the latter, students should take greater pains than their predecessors to distinguish relevant from irrelevant behavior.[35]

6.6 Indirect Effects of Behavior

There is a great difference between (a) stubborn adherence to an idea which is not tenable in face of contrary evidence, and (b) persevering with an hypothesis which is very difficult to demonstrate but against which there is no direct evidence.

Beveridge[36]

Observations in the last section were incidental to attempts to get stressed females to produce abnormal offspring. Having spent several years working out promising techniques, we applied them as follows. We first waited several months for colonies to reach a peak, then killed a sample of voles to see if they had enlarged adrenals or low glycogen reserves. Both indices went in the hoped-for direction; also, the young males had small testes.

Next, over the course of several weeks, we removed females just before parturition and let them have their litters in isolation. We then removed half of each litter and substituted the young of unstressed females. Luckily, we had a large enough breeding stock to provide litters of the right age. If E and C stand for experimental and control females, and e and c for their young, this design produced four sets of young animals, Ec, Ee, Cc, and Ce

$$E \begin{cases} c \\ e \end{cases} \quad C \begin{cases} c \\ e \end{cases}$$

The experimental mothers (E) weaned c and e young that were lighter in weight than those weaned by the controls; but the difference was slight, and merely showed that lactation of the E mothers had been affected in the colony or by the shock of isolation. This effect was irrelevant to the problem in the field, as there was no evidence of low weaning weights among young trapped in peak or declining populations (Gillian Godfrey's underweight nestlings never entered the traps). The crucial evidence concerned the Ce young crossfostered from experimental to control mothers. These young had had normal treatment as nestlings; so, if abnormal, must have acquired the condition before birth. Weaning weights were, as expected, similar for Ce and Cc young; the abnormal effects (in growth rate, maximum body weight, fertility, and viability) were not due until later. Making these observations took up much time and many cages, as we kept the animals and their controls for several months to compare their behavior. Nothing came of all this effort.

At the time we could not tell whether these results were contrary or merely negative results. If the former, the maternal stress hypothesis was wrong; if the latter, the stress was not equivalent to that believed to be affecting females in the wild. So the alternatives were to give up the hypothesis or keep trying to devise more realistic ways of producing stress. We chose the latter course, if only because we still had no alternative hypothesis.

If this had been our only line of attack we might still be blaming faulty techniques for our failure to produce effects *in utero*. We were rescued from such a bottomless pit by a wrong idea based on a misinterpretation of a faulty change in husbandry (a saga unfolded in Section 8.5). Trusting to serendipity to come to one's rescue is not the text-book way of conducting a scientific enquiry, but it's not to be sneezed at. Delbrück's "principle of limited sloppiness" has much to recommend it,[37] provided one is free enough from routine to be able to exploit the unexpected. "It's important", says Watson,[38] "to be slightly underemployed." A bureaucrat setting up a research lab would be unlikely to make this a top priority. Luckily, Charles's Bureau was anything but bureaucratic.

Controversies, 1951–1956

Politeness, Francis Crick said over the BBC at the time he got the Nobel Prize, is the poison of all good collaboration in science. The soul of collaboration is perfect candor, rudeness if need be. Its prerequisite, Crick said, is parity of standing in science, for if one figure is too much senior to the other, that's when the serpent politeness creeps in. A good scientist values criticism almost higher than friendship: no, in science criticism is the height and measure of friendship.

Judson[1]

If a house be divided against itself, that house cannot stand.

Matthew 3:25.

No Government can be long secure without a formidable opposition.

Disraeli[2]

7.1 A House Divided

In science as in politics, we need critics to keep us honest. Unlike the destabilizing effects predicted in the biblical maxim, divisive views in science are one of its strengths, as they ensure that its conclusions are subjected to intense scrutiny. Nevertheless, outsiders look on in amazement as grown men and women lose their cool over esoteric issues in population ecology. In acting thus, ecologists are behaving like other scientists, on whose sociobiology Hull[3] has cast his glittering eye. The following quotations and paraphrases give an idea of his views.

Scientists need not be unbiased, but different scientists must have different biases. In the resulting battles, including *ad hominem* attacks, scientists may seem to depart from a Platonic view of how they should behave. Indeed, some colleagues treat each other "with a steeliness the Borgias would have admired;" but this is "the way that all innovative scientists behave" and helps explain why science is so successful. "Scientists do not know what it is that they intended to say until they find out what other scientists think that they have said," Indeed, misconstrued criticism is especially helpful in making authors clarify their ideas. "Textbooks written by scientists are infamous for the degree to which they disguise the differences of opinion that character-ize scientists." Lecturers should make these conflicts known, however much their "students would prefer to memorize the truth, not a range of opinions."

Eating fruit from the tree of knowledge has always led to trouble, and the next four sections corroborate Hull's conclusions about the trouble one gets into through the pursuit of scientific knowledge.

7.2 An Editor's View

Spirits were broken, hearts were sickened, and authorship was cruelly discouraged by the savage and reckless condemnations passed . . .[4]

The account of my prewar work would have appeared in the *Journal of Animal Ecology* if Charles and I had not stepped down as editors.[5] But the new editor, H.C. Gilson, took a dim view of the paper and wanted it rewritten. Ten of the pages, he complained, seemed to be largely speculation. "Even if they were not, [he wrote] they would be too general in character to hang on to this paper." His comments confirmed the doubts I held at the time about the danger of speculating, especially as they were followed by phases such as the following: "All this is mere vague speculation . . . these *ex-cathedra* statements . . . if you will read the paper critically . . . infuriating to the reader . . . would you make up your mind . . . such indifferent photographs . . . much of interest and value pokes out of the paper like gleams of sun-shine in places."

I was more than somewhat upset—more than I should have been—and happened to come across the quotation above, which expressed the way I felt. This is probably the way most young authors feel when one of their brainchilden has been thrown to the lions of peer review. So it was some time before I could look objectively at the contrast between what I'd writ-ten—"needless to say I shall very gratefully welcome all your criticisms"—and the lack of gratitude I felt when they landed on my desk. The shock was the greater because I had been associated for 13 years with an editorial policy in which criticism had been tempered by a view epitomized by another quota-tion I had just come across: . . . personal opinions they remain, not truths to be imparted as such with the sureness of superior insight and knowl-edge."[6] Because reviews of new work are bound to reflect personal opinions, most editors now send manscripts to two or more reviewers, which dilutes the mental trauma that can be inflicted by a single devasting review from an anonymous source.

Criticism of one's manuscripts is a necessary and valuable part of one's education in science; but when my own turn came to act as referee, I tried to remember how easily I had been discouraged. Although sometimes tempted to use the sort of phrases leveled at my own manuscript, I try to prepare some delicately worded comments for the author, especially a young one, and a blunt set for the editor. And in the belief that a referee should be ready to put up with flak from disgruntled authors, I always sign my reviews.[7]

I next sought the opinion of a former colleague whose judgment I greatly respected (a respect confirmed in due course by the highest authorities) and later wrote: "I am grateful to Professor P.B. Medawar, F.R.S., for encouragement at a critical juncture." This was my understatement of the year. At this time, 1951, Sir Peter (as he became) had begun publishing his views about scientific method, one of which was that: "Today we think the imaginative element in science one of its chief glories."[8] His opinion of the need to speculate (though not necessarily of how I did it) restored my self-confidence. As well, he communicated the article to the Royal Society of London, in whose *Philosophical Transactions* it appeared after minor changes suggested by one referee, J.B.S. Haldane, and none by the other.[9] Although this was a prestigious place for it, the *Journal of Animal Ecology* would have brought it a wider circle of readers.

Anyone who proposes a new and doubtful hypothesis has a duty to test it; he should not leave it cluttering up the literature, and he alone may be willing to risk his career by working on it. Also, it's better to modify or reject it oneself instead of having it torn to shreds by one's critics. "Its [sic] lots of fun to blow bubbles, but it's wiser to prick them yourself before someone else tries to."[10] Little did I think I was starting a cycle of my own, in which blowing and pricking bubbles was to become the story of my life.

7.3 A Colleague's View

The Idols of the Cave are the idols of the individual man. For everyone. . . . has a cave or den of his own, which refracts and discolors the light of nature. . . . The Idols of the Cave take their rise in the peculiar constitution, mental or bodily, of each individual; and also in education, habit, and accident. Of this kind there is a great number and variety.

                                                            Bacon[11]

After the war, David Lack (Figure 7.1) became Director of the Edward Grey Institute of Field Ornithology and worked next door to the Bureau in the huts on the grounds of St. Hugh's College, later in the Botanic Garden, to which both institutions moved in 1952. David was the most vocal of my critics and the most lucid exponent of the ideas against which I struggled. "What determines population density? Animal ecologists debated this question in the postwar years with as much fervor as bishops at an Ecumenical Council and with as little thought of settling their differences experimentally. As one of the heretics, I was in schism with the orthodox doctrine of density dependence to which David Lack adhered."[12] As I criticize him later on, I here take the

Figure 7.1. David Lack around the time of the events narrated in Section 7.3.

*advice of Francis Crick[13] to begin by praising good aspects of the work
of someone you are about to rake over the scientific coals.*

*In the doubtful secrecy of our own circles, character assassination
went on with typical Oxford ebullience, exacerbated in 1951 by David's being elected ahead of Charles to fellowship in the Royal Society.
But once David and I had grown up enough to control our righteous
indignation at the enormity of the other's point of view, we became
quite chummy. For one thing, David invited me to be the guest speaker
at one of his annual conferences of zoology students. These conferences
were first rate and always a stimulating experience for budding orni-
thologists, besides being great fun. Apart from the work itself, there
was a light-hearted evening, for one of which David set an exam with
questions such as: "What kind of bird said 'Pieces of eight'? and 'What
bird 'ate a missionary on the plains of Timbuktoo, coat and bands and
hymn-book too'?"[14] His final act of courtesy was to arrange a farewell
tea in the Edward Grey Institute, making the unprecedented concession
of allowing his guests to smoke. This went over well with Helen, who
was, alas, a heavy smoker, a habit that undoubtedly contributed to her
fatal heart attack. In 1962 she and I enjoyed having David as a guest
in our new home in Vancouver and taking him on a field trip. I well
remember his delight at seeing the air full of bald eagles (19, if I remember rightly) on our ferry trip to Vancouver Island. He appreciated my
letting him comment on a highly critical review of his 1966 book before
I submitted it to Ecology.[15] He would, I think, have approved of the*

*following account of our differences, as he believed that expressing
things in black and white made it all the better to see them with.*

    *David was a man, "whose writings are gospel to some, anathema to
others and immensely stimulating to one and all,"[16] and it is unfortunate that in the small area in which our interests overlapped we judged
one another harshly. In a tribute to his many talents I reflected on how
our differences arose and how "a passionate devotion to one's own
point of view is probably essential to a scientist's need to ignore, explain
away, or otherwise cope with difficulties that would stop normal people
from even getting started."[17] I was glad to be consulted over a sympathetic obituary after his tragically early death in 1973.[18]*

While new ideas about the regulation of numbers were being explored at
the Bureau, they received a hostile reception from our next-door neighbor,
who, together with George Varley, was a firm believer in density dependence.
Only three factors, according to David Lack, can regulate the numbers of
natural populations (disease, predators or parasites, and food shortage),[19]
and he found it disturbing to think that fighting might also do the trick. Some
years later he made his beliefs perfectly clear: "the absence of field evidence
does not, and will not make the advocates of density-dependent regulation
change their minds."[20]

    As I considered David was putting back the clock by adhering so closely
to these ideas, there was little room for a meeting of minds, and we took pot-
shots at one another in various articles. We also locked horns at a meeting
at the Royal Geographical Society in London,[21] where ugly passions lurked
below the surface. All of this now seems childish; but on this occasion "scientific rationality [was] preserved in spite of such high emotions"[22]—an outcome that is less common than one might expect. Indeed,

> The polemics that make it into print are but the residue of the actual exchanges
> that go on at meetings, in private correspondence, and in manuscripts before
> they are sanitized by editors and referees. More than this, scientists cannot view
> things in any other way. Scientists involved in scientific disputes cannot help
> but see the disputes from their own perspectives. They would not hold the views
> they do if they did not think that they were right. Hence, any credence shown
> to alternative views must necessarily be wrong-headed.[23]

David was guilty, I thought, of turning a blind eye to the botanical studies
of Victor Summerhayes, which he failed even to mention in his 1954 book,[24]
stating that: "Unfortunately the food requirements of the cyclic rodents have
not been studied—an extraordinary omission." He also failed to answer other
objections to conventional views that Charles had given in *Voles, Mice and
Lemmings*. It's true that much of our prewar work was unpublished by the
time his book went to press, but he had seen the manuscript of my 1952
article, had attended our seminars, and should have consulted those of us
who had actually worked on cycles. In his book he acknowledged the help of
workers, who, "by their special knowledge corrected at least some of my

mistakes in subjects which I had not studied at first hand." Fear of being mistaken about cycles was not, however, a factor in David's choice of consultants.

David had several objections to the idea that crowding was what stopped voles from increasing. He pointed out that crashes start from different levels, instead of at fixed population densities;[25] that crowding does not explain why numbers continue to fall when the animals are no longer crowded; that harmful effects were unlikely to be passed maternally to later generations; that the regularity of the cycles was not explained; and that, according to this scheme, fighting was purely deleterious and had no survival value.[26]

David had a good point when he wrote that if fighting was merely a result of crowding "one would have expected the rodents to have evolved adaptations to counteract it."[27] C.D. Darlington also suggested I might be observing the effects of differential selection,[28] but I was too deeply bemused by my belief in maternal stress to pay it attention. Other geneticists, moreover, were offering discouraging advice.

David was right about the implausibility of my maternal stress explanation[29] and the unlikelihood that the harmful effects of strife "could be passed to the next generation (and to several generations in the varying hare)." He was also right about the need to explain fighting in terms of its survival value; but at the time I had no worthwhile ideas. (Later on I suggested that natural selection would favor genotypes that have a worse effect on their neighbors than vice versa. The idea was not new and was taken up again later; but theoreticians don't think much of it.[30]) David was also right in saying that "the basic mortality factors concerned may well be the same as in other and noncyclic animals,"[31] but was wrong in supposing I thought otherwise. We agreed in taking a unified view of cyclic and noncyclic populations; we differed because David thought the necessary change was in the severity of the mortality factors, whereas I thought it was in susceptibility to those same factors.

Had David consulted us, we could have given him the evidence against his predator-prey-like views. We could also have shown him 10 years of unpublished data on fluctuations, which might have disabused him of the idea that regularity is as common as he imagined. Furthermore, he missed the point that behavior and its effects rather than mere crowding were the variables supposed to initiate and prolong the crash, and that these variables are unpredictable from density alone.

It has been said that "the character of a scientific revolution depends in part on its opposition",[32] and the same goes for lesser shifts of perspective.[33] I benefitted immensely from my jousts with David, partly because of the soundness of his views on natural selection, but primarily because I realized that to convince the opposition I would have to present a stronger case. I'm still working on that one.

7.4 The Australian Schismatics

Doubt is the road to enquiry; by enquiry we learn the truth.
Peter Abelard (1079–1142)[34]

Every great advance in natural knowledge has involved the absolute
rejection of authority, the cherishing of the keenest scepticism, the
annihilation of the spirit of blind faith.
T.H. Huxley (1825–1895)[35]

In the same year, 1954, that David Lack produced his views on the regulation
of numbers, Andrewartha and Birch[36] proclaimed that regulation is a myth,
that "there is no need to attach any special importance to 'density-dependent
factors,' " and, instead, that populations alternate between realizing their intrinsic rate of natural increase and being struck down by adverse components
of environment. Their ideas received a cold review from Charles,[37] though he
gave the authors credit for ". . . so much that is useful and provocative to
thought . . . sincerely tackled and expounded. . . ."

> This large monograph begins by rejecting the relevance of any study of whole
> animal communities or ecosystems to the understanding of distribution and
> numbers of animals. It then proceeds to take a firm stand against the general
> use of mathematical models and, in particular, of ideas connected with biotic
> relationships and density-dependent processes between populations, as explana-
> tions of events in Nature. The authors believe that the regulation of numbers
> does not exist in the sense hitherto examined in theoretical and laboratory pop-
> ulation ecology. . . .
>
> Most of this section ['Analysis of Environment'] about population relation-
> ships is written in a highly critical, often polemical, manner, and it forms a
> sustained and oddly contemptuous attack on what the authors regard as the
> dogma of the regulation of numbers through varying density-dependent pro-
> cesses based proximately upon biotic links. It does not seem likely that the
> reader will be entirely impressed by the implication that nearly all previous
> research in the subject is irrelevant to the interpretation of field population
> changes, or that the study of simplified mathematical models and their testing
> in the laboratory is misleading. . . .
>
> There are some strange remarks on important topics: for example, that
> something cannot be a general theory unless it describes a substantial body of
> empirical facts . . . and that only individuals, not populations, can be said to
> have an environment. . . .
>
> I cannot believe that the natural communities of the world will be found to
> have evolved solely "by guess and by chaos," or that the inter-relations of ani-
> mal populations have not introduced a gradual order into the structure of ani-
> mal communities. . . .

This review is the best evidence I have of Charles's feelings towards those
who questioned the value of community studies, the prevailing ideas about
density dependence, and the use (or abuse) of mathematics. It helps explain

some of the trouble I got into through propounding similar iconoclastic views. Though I did not go along with the idea that populations are unregulated, I agreed with Andrewartha and Birch in their scepticism about current approaches to population and community ecology, including the Wytham Ecological Survey, to which Charles devoted so much of his energy after the war. I particularly liked the authors' conclusion that only individuals can be said to have an environment. I also agreed with the restricted use of 'theory.' In common parlance 'theory' and 'bright idea' are often used synonymously, whereas in science "A theory is the whole system of statements comprising hypotheses and the statements they entail."[38]

I myself thought the authors' most valuable contribution was their integration of experimental and field ecology and contrasted their approach with that of David Lack, who seemed to share with certain naturalists "a mistrust or misunderstanding of experimental methods." I wrote that:

> On the grounds that the animals are kept under artificial conditions, Lack feels justified in rejecting all but a few selected results from the work on experimental populations. (It is fortunate that Torricelli did not abhor a vacuum and that Darwin was willing to learn from variation under domestication.)[39]

In contrast to their own experimental outlook, Andrewartha and Birch based many of their beliefs on the frequent association between changes in weather and population density, despite the fact that these associations were mostly *a posteriori* descriptions. But I thought their challenge to the prevailing density-dependent dogma (as they termed it) wholly salutary—a typical reaction of one who finds support for his own ideas.

7.5 The Director's View

Anybody who still carries in his mind the picture of the cold, unemotional, impersonal scientist should [forget it] . . . the blood of scientists runs hot no less frequently than that of others, and probably rather more so. Most scientists are deeply interested in their work and nobody can watch the reaction of others to what he is deeply interested in, and to which he had given of his best, without becoming emotionally involved.

Bondi[40]

A theory which is held as a faith may stifle progressive research and turn fresh facts into personal insults.

Elton[41]

It might be thought from Peter Crowcroft's book[42] that nothing but sweetness and light prevailed at the Bureau. But we were a bunch of individualists whose critical views were not exclusively directed against the Cambridge Mafia (David Lack at the Edward Grey Institute and George Varley at the Hope Department of Entomology were graduates from 'the other place'). We had our own internal differences, most of which were aired at seminars and during morning and afternoon breaks for tea or coffee. Our differences were useful in showing students that we could disagree violently while remaining on civilized terms. Not all institutions of learning are that lucky; communication is sometimes difficult among supposedly rational academics, who tend to mistake themselves for prima donnas or Heldentenors; and in another job I might have been out on my ear for being as critical of my boss as I sometimes was of Charles. It is a tribute to him that he put up with my criticisms and a failing on my part that it was years before I realized that on some issues he was more sensitive than the rest of us. There are few documented accounts of the bruised feelings associated with these clashes of opinion. I am lucky enough, however, to have kept a letter from Charles that sets forth my main deficiencies and the issues that divided us. I regard this letter as a document of historical importance as confirming Crick's view that "in science criticism is the height and measure of friendship." By this criterion Charles and I were exceptionally good friends, as indeed we were.

Charles once told me that if you were going to have a row with someone, you should hold your fire until you could have a real showdown. He demonstrated this technique by bottling up his displeasure over certain early incidents until after a chat in June 1956 about the lynx cycle. Unfortunately, I have no note of what I said, nor which of Charles's ideas I was challenging. I suspect it was his views on synchrony, about which I had made the following remarks at Munich: "On present evidence, in fact, the continent-wide synchrony of the cycles in the 1920s was a chance result, and the length of the cycles is different in different regions."[43] For whatever reason I found myself on the receiving end of one of those Jovian thunderbolts this mild-mannered man occasionally hurled at people who had upset him. Part of the opening salvo went as follows:

In the last few years you have shown a strong tendency to do what —— does, select the facts that fit and under-rate those that don't. This in spite of your declared policy of doing (what every sensible scientist does, without having to give it any very technical language) your best to see if exceptions to the theory disagree with the main idea. . . . [As a result] you will more and more sit back into a team-world of your own, with people round you that agree with everything you say.

Here, Charles got carried away. My closest colleagues at the time, Helen, George, John Clarke, David Jenkins, and (later) Janet Newson, though not always successful, did their best to prevent me from making a fool of myself.

To quote this letter in full would give the false impression that our relations were soured by our differences; but it's worth quoting the juicier passages, because they show, first, how passionately scientists feel about their work, second, how difficult it is, even under conditions as good as those at the Bureau, for colleagues to understand one another's point of view, and third, how valuable for scientific integrity is the role of unbridled criticism.

Charles and I had worked harmoniously together on voles before the war and on rodent control during the war but thereafter went our separate ways. Unlike George, Charles no longer understood the shift in emphasis in the vole work, and I was indifferent to the Wytham Ecological Survey ("It is 15 years since you asked me what I am doing or trying to do in ecology"). I believed that without an experimental approach, such as that of Bill Fager,[44] ecology was doomed to remain glorified nature study. So, in pronouncing my views about the Wytham Ecological Survey, I was so tactless as to say it looked like 19th-century natural history. For once, David Lack and I were on the same side of the fence; David called it natural history on punch cards. As already shown, I was not alone in my views; and some years earlier O.W. Richards had been equally sceptical "of the value of applying the concept of the community to animals" and had become "convinced that the key to progress was in the reductionist approach, investigating the population dynamics of individual species"—views that had created great tensions.[45]

Even if such views are justifiable, Charles's best ideas occurred to him, so he told me, during the mechanical work of collecting, mounting, and classifying his collection of invertebrates. It provided the creative atmosphere that von Helmholtz found "during the slow ascent of wooded hills on a sunny day." [46] In response to my thoughtless remarks, Charles wrote: "You already have a fairly damnatory attitude towards people (including myself) . . . nor do you have a clue to what the effect of your speech or writing is upon peoples' feelings and attitudes. You are undoubtedly a discouraging colleague."

It was not that I despised natural history, which is an essential precursor to science, but I claimed it should not be regarded as science until its irrefutable statements about *some* characteristics had been replaced by testable statements about *all or a stated proportion* of them. "Scientific natural history," I claimed, was an oxymoron, a claim that was poorly received:

> Owing to your powerful introverted mind, with its capacity for sustained logical thought, you will feel that no one else could possibly think about the subject to any purpose, without the same mind and the same enormous amount of concentrated thought and research. You will feel you have become the only person in the world whose ideas on the subject bear examination. And you already have a fairly damnatory attitude towards people (including myself) who do have other ideas, while respecting yours. And as you don't read much,[47] you will overlook your predecessors, or analyze their sentences instead of their paragraphs.

My attitude had been shaped by the surprise of finding that evidence for Charles's epidemic hypothesis, which as an undergraduate I had accepted as gospel, was built on faulty inferences. To find out what had gone wrong, I took the advice of Brian Farrell in the Sub-department of Philosophy and read Cohen and Nagel's *Introduction to Logic and Scientific Method*.[48] This book more than any I've read before or since has shaped my views about scientific method, though the writings of Bernard, Whitehead, Beveridge, Popper, and Medawar are runners up. The debt I owe to these and other writers will be obvious from the quotations throughout the book. Like a religious convert, I took upon myself the evangelical mission of proclaiming the faith that "the separation of relevant from irrelevant factors is the beginning of knowledge."

The problem of 'relevance' led to a misunderstanding that's worth analyzing. Charles had written: "Your own ideas are mainly consistent, *within the field you are prepared to accept as relevant.* To you, predators are not relevant, so let's forget them . . ." Predators are an obviously important element in the dynamics of most populations, but the meaning of "important" is equivocal, whereas the meaning of "relevant" can be decided experimentally.[49] Charles himself had pointed out in 1942 that the vole system was less dependent on other animals than we had first supposed, and that a vole taken by a predator might have died anyway.[50] By 1956, however, he had reverted to what Errington has described as "the spurious logic of reckoning population effects of predation . . . merely in terms of losses to one predator or another."[51]

Errington mentioned "the usual human tendencies to overestimate the population effects of conspicuous or demonstrably heavy predation" and stated that surplus animals, if not killed by "locally active native predators," might be removed by motor cars or farm dogs. The effect of intraspecific strife was, thus, to produce "a biological surplus, largely doomed through one medium or another." In believing that populations can limit their own increase and that causes of death will vary locally, I saw eye to eye with Paul Errington.

We have already seen that Charles regarded the criticisms of Andrewartha and Birch as "oddly contemptuous," and he had a similar reaction to some

of my remarks. Referring to our chat about lynx, Charles wrote: "Your final remark yesterday about 'so-called' being the only word needing alteration in your speech [at Munich[52]] was so funny that I could not even feel it to be insulting—which it was." He himself, he wrote, was a person who "hardly ever expects to be permanently right (except about the lynx cycle, and I will fight all comers for a very small purse on that)." "It is inevitable," he wrote in another connection, "that I shall not hit the truth exactly (which is an admission you would never make about anything you wrote!)." I was also taken to task for some editorial surgery I had performed in editing *Control of Rats and Mice:*

> Having created [a] mental picture which satisfies you as a scientist, you go on to incorporate in it the dynamic force of your own personality, with pride and ambition attached, and the result is that it appears practically treasonable for anyone to suggest altering a word in one of your carefully prepared papers! But it does not appear strange to you that you should alter words in other peoples' carefully prepared MSS.

In helping edit the *Journal of Animal Ecology*, I had learned from Charles how to deal critically with manuscripts. The trouble with critical disciples, however, is that they don't know where to draw the line, and he who sows the wind may reap the whirlwind. I was also apparently guilty of having insulted an intelligent audience at the Royal Geographical Society "by doing tricks with woolly rabbits and speeches from the audience;" so Charles vowed never to attend any future talks of mine.

To explain what had gone wrong at this meeting,[53] I must confess to a misguided attempt to entertain as well as enlighten the audience. In the belief that scientific talks need to be leavened with a modicum of humor, I had prepared a tapestry showing cotton-wool snowshoe hares running down their population curve against a black cloth background. The audience was not amused, but I had no idea I had given offense. I have since rationed my frivolity more carefully, though sticking to the view that most scientific talks need pepping up. I have forgotten what was wrong about "speeches from the audience," but think I held a prepared dialogue with Helen, to show what she had contributed.

Charles advised me to sacrifice some of my "interest and ambition about getting on with only" my own work,[54] as I would otherwise "be just as much a dead loss to the B.A.P. socially and intellectually as ——— has been." The letter ended by urging me to "realize that the large problems can only get solved by using all kinds of minds to work on them" (a case, surely, of the pot calling the kettle black). In reply to my dressing-down, I wrote as follows:

> My dear Charles,
>
> I am very sorry to know how you feel towards me. I realize that a lot of the trouble comes from the way I express myself on occasion and I will try to correct this defect at least. The other difficulties may take time to iron out but I've no doubt we shall overcome them between us.
>
> Let me say how much I appreciate the way you have always supported and encouraged me in spite of your profound misgivings about my points of view.

The trouble with the policy of unbridled criticism is its asymmetry: it's more blissful to give than to receive. But however hard on the recipient, the freedom with which criticisms were exchanged between people at all levels of seniority was one of the reasons for the high standards for which the Bureau was famous. In the present instance, life returned to normal after an upsetting day or two and left no hard feelings on either side (nor, alas, left a record of what the fuss was about). As a goodwill gesture I collected some insects in Wytham but unfortunately mixed predators and prey in the same vials, so my contribution to the Wytham Ecological Survey was not all it might have been.

A saying often used by one U.S. President is that "if you can't stand the heat you better get out of the kitchen." In the words of another President, "let me make it perfectly clear" that my getting out 5 years later had nothing to do with this clash of opinions. Anyone who can't take criticism should forget about a career in science. Charles and I parted with regret, respect, and affection.

Varying the Circumstances, 1952–1959

8.1 Game Birds

There can be no true physical science which looks first to mathematics for the provision of a conceptual model. Such a procedure is to repeat the errors of the logicians of the middle ages.

Whitehead[1]

For someone studying behavior, the reclusive vole was a poor choice of animal; despite its merits, it was difficult to observe in nature. Partridges are easier to watch though less easy to experiment with. Having both species under study at Oxford gave us the best of both worlds. Partridges were one of the game animals whose fluctuations in numbers were studied by Doug Middleton early in the history of the Bureau.[2] On some estates partridges are managed by gamekeepers, and records of numbers shot have been kept since the 18th century. Shooting is a different sport in England from hunting game birds in North America. Local people are hired as beaters to drive the partridges over the shooters, who usually have two guns and a man to load them. After the massacre, the birds are laid out in rows while well-dressed ladies and gentlemen sit down to an al fresco lunch spread on trestle tables laid with white linen and groaning with food and drink.

Surplus birds are sold to a local dealer, who pays a smaller price for adults than for young. One of Helen's jobs was to attend these shoots. Having learned how to tell young from old, she discovered that the dealer had his own criteria, which, not surprisingly, skewed the age distribution towards the less valuable old birds. Besides going to shoots, Helen helped assemble game bag records and study the food of partridge chicks.[3] (Friends found it odd that she was dissecting a batch of chicks until an hour before our wedding. First things first, was her motto.) Besides analyzing game bag records, Doug studied the natural history of the partridge and believed that numbers went down especially fast when the birds dispersed at pairing in late winter.[4]

Unlike the figures for numbers of trapped lynx, those for shot partridges are irregular and, according to Pat Moran, not oscillatory.[5] Mathematicians naturally expect God's creatures to behave according to the Almighty's mathematical wizardry. But mathematical criteria are not necessarily the best for

recognizing similarities among biological phenomena and indeed may empha-
size differences. Luckily, there were few mathematicians around in the early
days. Nor had David E. Davis handed down his verdict that "The emphasis
on cycles . . . has had a disastrous effect on the analysis of regulation . . .
To refer to microtine and other populations as cyclic is a gross error because
it leads to the wrong formulation of the problem." [6] He does not say which
is the right formulation but gives a clue by saying that in cycles "one periodic
factor regulates the increases and decreases" and that this directs research
into looking for "the cause (singular) of the cycle." This was news to me.

Being ignorant of the error of their ways, prewar biologists went ahead and
assumed that, despite a lack of strict periodicity, fluctuations were more often
alike than unlike. After the war, however, David Lack wrote an indignant letter
to the editor of the *Journal of Animal Ecology*[7] complaining almost of libel because Gordon Williams[8] had misquoted him as saying 'cycle,' when he (David)
had been careful to say 'fluctuation.' He had a point, of course, because among
mobile populations living in the midst of agricultural change and being managed, shot, subject to economic trends and wartime changes to human targets,
not all fluctuations are likely to belong to a single class of events.

My own interest in game birds stemmed, first, from the belief that fluctuations in their numbers could often be explained in the same way as for voles.
Of the criteria developed for recognizing cyclic declines in voles, one of the
most characteristic is the continuing decline or refractory period of low numbers well after crowding has ceased. As many partridge populations (or rather
the numbers shot) also went down for 3–4 years after a peak, it seemed likely
that some, at least, might be regarded as cyclic. Second, experience with voles
had left me unimpressed with beliefs based on correlations between disease
and population declines.

Contemporary wisdom dated from the findings in 1911 of a Committee of
Inquiry on Grouse Disease[9] that reductions in numbers of red grouse were due
to the effects of a nematode upon the adults and coccidiosis upon the young.
But the Committee also realized that infected populations did not always decrease and that the spread of parasites alone did not account for a change in
mortality rate. Still earlier investigators, moreover, had observed that birds often died in apparently good health with no abnormal conditions that could obviously be associated with the losses of chicks; and a contemporary author had
also shown that strongylosis was not always associated with the die-offs.

A few workers had therefore postulated that some antecedent condition
was responsible for the unexplained losses; the earliest suggestion, in 1868,
stated that mortality was due to a "state transmitted from parent to offspring, and predisposing the young to suffer from influences such as severity
of seasons or temporary scarcity of food, which under other circumstances
they would have resisted successfully." A later author had suggested there
was "a general weakness of the progeny . . . inevitable even before laying."

These findings agreed well with my own ideas and persuaded me that the
regulation of numbers in game birds was due to changes in physiological condition. At the same time, the findings of the 1911 report agreed with David

Lack's ideas and persuaded him that regulation of the numbers of red grouse was due to starvation and parasitism.[10] "So, in another tribute to human ingenuity, each used the same data to support an opposite point of view."[11]

According to earlier ideas, the increase in susceptibility was due to unfavorable conditions in weather and food; and some support was naturally available, as the British weather is likely to be abnormal at some critical time in an animal's life. According to my ideas, the increase in susceptibility was due to unfavorable social conditions among peak populations; and from the current literature it seemed that mortality factors varied so much from species to species as to bear no general relation to population declines. I was therefore delighted when my next D.Phil. student was keen to study the regulation of numbers in terms of the behavior of the birds themselves rather than in terms of predators, parasites, disease, or food supply.

8.2 Partridge Behavior

For the confirmation of a hypothesis . . . the greater the variety,
the stronger the resulting support.

Hempel[12]

David Jenkins, a graduate of Cambridge University, came to the Bureau in 1952 and obtained his D.Phil. in 1956 and an Oxford D.Sc. in 1982. Working on his own, he did a remarkable study of the population dynamics and behavior of a population of partridges. This work has not, I think, received the recognition it deserves, though the methods he worked out contributed greatly to the success of a later study of red grouse.[13] Environmental conditions have since changed for the worse with the introduction of pesticides; and partridges may now be too scarce in many places to show the self-regulatory behavior they showed in 1952–1956.

Finding a site for David Jenkins's partridge study turned out to be easier than expected, thanks to good relations between Doug Middleton and Sir J. Arthur Rank (later Lord Rank), and to the head gamekeeper having forgiven Doug for his breach of etiquette in the van. Doug and I met Sir Arthur to ask for permission to study partridges on a few acres of his estate, 60 miles south of Oxford. I gave a brief account of the unsatisfactory nature of current population theory, my latest ideas about vole cycles, their possible relevance to other organisms, and the chance that they might lead to better management of partridge stocks. The great man listened with apparently rapt attention, one hand cupped behind his left ear, and as soon as I shut up delivered the following succinct verdict: "Professor, I don't understand a word you've said, but if you want a few acres you can have the Borough Beat." The Borough Beat was 640 acres, far more than I had dared expect.

Not only did Sir Arthur turn David loose on such a large part of his estate, but he provided a cottage, a Jeep, unlimited petrol, the help of a gamekeeper, and freedom to use whatever methods he chose. As a result of this generosity David was able to live among his birds for months at a time. Unlike less fortunate students, Oxford men and women can get down right away to research, unhampered by courses, language requirements, and comprehensive exams.

The partridge work was of two kinds: determining numbers and quantifying behavior.[14] In the 4 years (1952–1956) numbers varied greatly at the end of the breeding seasons (Figure 8.1), although they were much the same in midwinter and spring. Adjustments between autumn and spring took the form of a steady decline at first, followed by a sudden drop after mid-winter, when pairs began to form. Over half these losses, largely through dispersal, took place in the 3 weeks before February 8, 1953 and in 9 days in February 1954 (Figure 8.2). And in the next year over three-quarters of the total late-winter loss occurred between mid-December and March 3. Before January and February coveys avoided one another, but later on

> Coveys as a whole became much more pugnacious, and would usually join any fighting or chasing in the vicinity, even if it were outside their normal home range. On occasion there would be as many as 50 birds chasing each other indiscriminately within a small area. Cocks would chase and fight hens; brother would chase brother; and sister sister. Unmated cocks would viciously peck potential mates, and parents would attack their offspring. There was complete anarchy. Such disorders occurred frequently, and, in various stages of complexity, sometimes lasted for hours. When the coveys reassembled it was not unusual to see aggression between their members.[15]

The number of surplus birds eliminated between autumn and spring depended mainly on the number of chicks surviving the first month after hatching: the better the survival, the heavier the losses later on. Many of the early deaths could be attributed to bad weather, resulting lack of food, disease,

Figure 8.1. Number of partridges present during peaks in autumn 1953–1955 and low points in spring, when numbers were the same regardless of numbers in the previous autumn. Modified from Jenkins (1961a: Table 2 and Figure 2) to show acceleration of losses in spring.

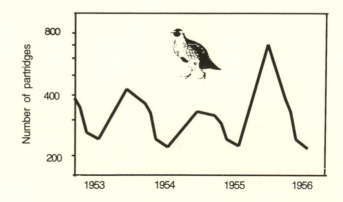

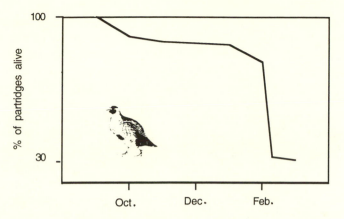

Figure 8.2. Percent of partridge population alive between September and March 1953–1954. From Jenkins (1961a, Tables 2 and 4 and p. 243).

and other obvious mortality factors; but as in past studies, many occurred inexplicably under apparently favorable conditions.

David spent most of his time studying the behavior of his birds, many of which were individually marked. He grouped activities in two categories: those resulting from intraspecific interactions (fighting, chasing, threat, etc.) and those occurring independently of interactions with strangers (courting, feeding, etc.).

In the first year of the study David had had no success in relating differences in these activities to differences in population density alone; some other variable was also relevant. To get an objective measure of differences in cover, David and Denys Kempson devised a photographic technique. Figure 8.3 shows a number of partridge models as seen from partridge-eye-level. Pairs that were alone or in good cover (0–3 models visible) spent most of their time undisturbed except by calls from distant neighbors. Pairs in over-grazed meadows (all six models visible) or near other pairs or unmated cocks were disturbed more often. Thus, the proportion of time spent in different activities depended not merely on population density but on habitat type.

This thumbnail sketch covers only those of David's results that concern my own vested interest, which was mainly to know, by the direct observation denied me on voles, whether partridges have a form of behavior that sets limits to their breeding density. Apparently they do; though not territorial, they space themselves out by avoiding other birds. A second interest was to know whether unexplained deaths of chicks were related to maternal stress. David did indeed find "an inverse correlation between interaction among adults in the late winter and subsequent chick survival," but neither of us regarded these results as more than encouraging. And as there was no cyclic decline, David was unable to throw light on the critical point of how juvenile survival differed between that expected during a cycle and that observed in stationary or increasing populations.

A cyclic decline would have made it worthwhile repeating an experiment

Figure 8.3. Method of assessing density of cover. In poor cover all six models representing partridges are seen from partridge-eye-level. The better the cover the fewer the number seen. Spacing: 10 yards. From Jenkins (1961b).

for which the argument went as follows.[16] If chick survival varies with population phase, the way to find out is to switch eggs between hens from different types of population. Knowing that chick survival had been poor in the years before he began his study, David collected eggs from areas where chick survival had been good, incubated them under domestic hens, and "allowed wild partridges to rear [the chicks] side by side with others that had kept their own eggs. Twenty-two clutches were exchanged. The experiment was a failure, because all the eggs hatched during a period of unusually heavy rain in June 1953, and introduced and indigenous chicks died without discrimination." Another interpretation might have been that the experiment succeeded: namely, in showing that there was no qualitative difference between chicks. Luckily, we didn't think of that—we might have been discouraged.

This experimental design was similar to but neater and more practicable than the fostering experiments I was doing at the same time on voles. It was many years before anything similar was carried out; then it was on an insect that was even better for this type of experiment.[17]

From my point of view this study was a good test of the hypothesis that, in favorable environments, all species can regulate their own numbers and that the process is unpredictable from density alone. It was unfortunate that the population experienced only seasonal declines. On the other hand, there was no reason to expect anything else, as this population was invaded by birds from outside. Thus, the *ceteris paribus* clause was unsatisfied that the delayed consequences of self-regulation may be undetectable in environments swamped by immigrants from populations in other phases.

The sharp change in survival, accompanied by a change in behavior, made it plausible to think that partridges can set an upper limit to their breeding density. And it was not long before David and Adam Watson, studying red grouse,[18] confirmed this role for behavior (Figure 8.4).

David's conclusion that his egg-swapping experiment was a failure was based on his belief that the weather was so bad that no young chicks, however healthy, could have have survived it. Bad luck in this case, bad planning or unforeseeable circumstances in other experiments, can spell failure through preventing one from reaching a conclusion one way or the other. In all other cases an experiment that goes against one's expectations must be judged at least as successful as one that supports it[19]— more so as a rule. Judged by the latter criterion, my career has been a great success.

8.3 Vole Adrenals

We should quite often have occasion to say "I used to think that once, but now I have come to hold a rather different opinion." People who never say as much are either ineffectual or dangerous.

Medawar[20]

From 1949 to 1956 we thought that weighing adrenals and other organs would corroborate the maternal stress hypothesis. On each visit to Lake Vyrnwy we collected animals in snap-traps and froze them in dry ice. In the following weeks Helen thawed out a daily quota of corpses and weighed them and their adrenals, thymus, spleen, liver, testes, seminal vesicles, and

Figure 8.4. Tracks made in the snow by red grouse during a boundary dispute. From Jenkins and Watson (1970); original photo courtesy of Adam Watson.

embryos, counted corpora lutea and placental scars, and noted the shape of the skull, which changes with age. She also dissected lab-bred animals for comparison. Knowing from John Clarke how fighting affects the weight of some of these organs, we expected to find differences between increasing, peak, and declining populations, which is indeed what Helen found[21] on one of the two areas we were studying.

Figure 8.5 shows that between 1953 and 1956 the Old Road population went through all four phases—decline, low, increase, and peak. As expected, adrenal weights were highest during the decline, lowest during the increase, and higher again during the peak (Figure 8.6). By themselves, these data for the Old Road population would have confirmed our preconceived notions. Luckily, we had another string to our bow. The Forestry Commission population was out of phase with that on the Old Road, but adrenal weights on both areas went up and down at the same time instead of with the phases of the populations. So differences between years must have been due, not to differences in phase, but to differences in weather or other local conditions. Comparative methods had once again proved useful in eliminating a mere correlation.

Within each year adrenal weights tended to be lower in March and April than in May and June and to be lower again thereafter. This seasonal rise and fall was apparent "in all categories (10) which were represented in sufficient numbers for a test to be made, i.e. for mature and immature animals of both sexes." Adrenal weights were higher in pregnant or lactating than in nulliparous females, results that were consistent with lab observations.[22]

Figure 8.5. Population trends according to a trap index, together with mean body weights of male voles for June 1952, March 1953–1956, and May 1957–1960. The Forestry Commission area (FC) was first studied in 1953; the Old Road area (OR) was studied in 1952 by live-trapping only. From Chitty and Chitty (1962a,b) and Newson and Chitty (1962). FC---; OR—

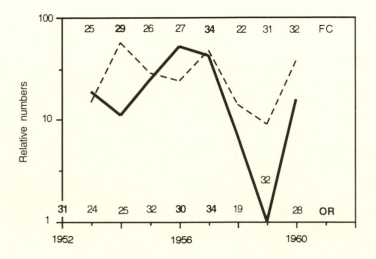

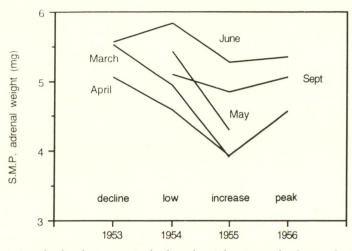

Figure 8.6. Standardized mean paired adrenal weights (mg) of voles on the Old Road during 4 phases of a cycle in numbers (see Figure 8.5, OR. From H. Chitty, 1961).

Thus, if adrenal weight is an index of stress, life is less stressful early and late in the breeding season than in June, but no less stressful at low than at high densities. Indeed, the highest adrenal weights were among females in low and expanding populations. Once again, population density had been found wanting as a predictor of the state of a population.

After the first 2 years' discouraging results we debated whether to keep the study going. It's hard to decide whether to stop an apparently unprofitable line of work or carry on hoping for the best. A good set of data may be useful, perhaps in unforeseen ways. Also, having invested time, effort, and money, research workers are tempted to continue fishing rather than cut bait. They may at least get a publication out of it. Luckily, none of us at the Bureau suffered from the peril of having to publish papers whose only justification was that we had done a lot of work. We were therefore able to jettison years of unprofitable data that might otherwise have added further clutter to the literature. Nor did we feel obliged to claim joint authorship with students whose work we had merely suggested, supervised, and edited. Today's competition for funds, which often depends less on the quality than the quantity of one's publications, can perhaps be blamed for the frequent demise of this policy. A one-line vote of thanks is less impressive than joint authorship, and producing four short papers does more for one's paper count than putting them all together.

By the end of Helen's long and demanding study of organ weights we had rejected our original expectations and, instead, had become all fired up over her discovery that spleens were grossly enlarged, a discovery that had consequences of a wholly unexpected kind.

8.4 Vole Spleens

A new discovery is usually overstated and very often formulated
into a hypothesis with too much detail that wears off later.

Schrödinger[23]

Helen's measurements of the spleen could have been made into a good descriptive paper if events, moving faster than her pen, had not made her results
obsolete. The only published record of her work is in a paper of mine given
at Cold Spring Harbor in 1957.[24] After reviewing our postwar work, I listed
three tasks: (1) to discover an abnormal condition associated with a decline
in numbers, (2) to produce the same condition experimentally, and (3) to
examine a wide enough variety of instances to show that the condition was
invariably associated with cyclic declines. In my heart of hearts I thought (1)
and (2) had been satisfied, but I'm happy to say that I concealed this convic-
tion and emphasized that (3) had not. Then followed the argument, which
went as follows: (a) Among 80 animals maintained in the lab from 1952–
1954, 70% had spleens weighing less than 31.5 mg, and none had spleens
over 100 mg. Among field animals, spleens were seldom less than 100 mg,
some weighed over 500 mg, and one weighed over 1 g. (b) Since October
1955 many voles, living in the lab under the same conditions as in former
years, had been found to have spleens far heavier than anything previously
observed, even among stressed voles. (c) The only factor known to be associated with this splenic hypertrophy was that experimental animals and their
progeny had, since September 1954, been freely incorporated into our breeding pairs in the hope that some such change might become apparent.

Although hedged around by warnings that they were incompletely analyzed, perhaps of local significance only, and depended on unsubstantiated
assumptions, these results were sufficiently striking to lead me to give a talk
I wish I'd never given. Stripped of its cautionary rhetoric, the paper amounted
to yet another premature claim to the discovery of the mysterious condition
responsible for cyclic declines. I'd fallen into the age-old trap of letting enthusiasm for a preconceived notion get the better of scientific caution. As it happened, however, the only thing people remembered about this talk was my
announcement that Helen, who had brought along additional graphs, had
some curves she would be happy to show to anyone interested.

8.5 Hemolytic Anemia

Even mistaken hypotheses and theories are of use in leading to
discoveries . . . we might still cite men of science who make great
discoveries by relying on false theories. It seems, indeed, a necessary weakness of our mind to be able to reach truth only across a
multitude of errors and obstacles.

Bernard[25]

Wrong hypotheses, rightly worked from, have produced more
useful results than unguided observation.

De Morgan[26]

In mitigation of the way I went off the deep end over these enlarged spleens,
a tolerant reader will understand my frustration at having bumped my head
against so many brick walls. I was especially ashamed at having uncritically
accepted the shock disease story and, as part of my talk at Cold Spring Harbor,[27] explained why I thought it wrong. David E. Davis, who was in the
audience, telephoned his colleague, John Christian, who arrived a day or so
later to confront me over this attack on work that had played so important
a part in forming his own ideas. Though united in opposition to conventional
views and agreeing on a role for behavior, he and I disagreed over the reason
why its effects persisted into the low phase. Whereas he had adopted my
original view that these were maternal effects,[28] I now believed that they were
more likely due to natural selection. I did not deny that crowding might have
adverse maternal effects, but now assumed that the more powerful they were,
the more likely they were to be counteracted by natural selection. As both
hypotheses have been fruitful and are not incompatible, we have gone amicably on our separate ways.

The first 10 years work after the war had encouraged me to think that the
first part of the maternal stress hypothesis might be true, that is, that voles
are *capable* of limiting their own numbers through social antagonism
(whether or not they always do so). But there was no empirical support for
the second part, namely, that stressed females give rise to at least two nonviable generations; nor was there any theory to lend confidence to this interpretation. (Theory is indispensable to discipline mere data, as are data to discipline mere theory.) It nevertheless seemed best to muddle on in hopes that
something would turn up, which it did.

In fact, five things turned up. The first was Helen's discovery of splenic
hypertrophy among wild voles, the second was the incidence of an apparently
identical condition among the laboratory voles looked after by Ellen Phipps,
the third was Janet Newson's coming across a paper on hereditary spherocy-
tocis, the fourth was an invitation to a conference at Cold Spring Harbor,
and the fifth was something whose relevance escaped me until too late.

Being unaware of the latter, I advanced the following argument.[29] (1) The
spleen becomes slightly enlarged under stress; in the field it was grossly enlarged, presumably due to even greater stress. (2) We had subjected one or
both members of most breeding pairs to stress; many of their descendants
now had large spleens. (3) The phenomenon of splenic hypertrophy in the
field was qualitatively identical to that in the lab. (4) We had recognized an
abnormal and apparently relevant condition in the field and also produced it
in the lab. (5) The condition was a heritable form of hemolytic anemia. (6)
Selective pressures explained the incidence and degree of spontaneous splenic
hypertrophy among the descendants of stressed animals in the lab and probably did so in nature.

At the time I was unaware of Murphy's sixth law, which states that "If everything seems to be going well you have obviously overlooked something," though Bacon had said the same thing 3 centuries before. My normal tendency to put off writing has probably saved me from publishing even wilder ideas. But this time the invitation to give a paper came before I'd had time to sort out right from wrong, and I experienced the double jeopardy of falling prey to my preconceptions and committing the fallacy of *post hoc, ergo propter hoc.*

Although George Leslie was scornful of my assumptions, I pushed ahead, all unconscious of my fate, having been blinded by an early decision on whether to keep stocks of stressed and unstressed animals from getting mixed together. I had decided that if our experimental methods were ineffective no harm would be done by pooling the stocks, and that if the methods were effective the problem would have been solved. Later developments saved me from complete disaster. First, Janet Newson and I demolished our own hypothesis in 2 years, and though publication took 4 years more, the interval was relatively short. For it is not the case in population ecology that the "process of invention, trial, and acceptance or rejection of the hypothesis goes on so rapidly that we cannot trace it in its successive steps."[30] Ecologists have the doubtful advantage of viewing the process in slow motion. Second, while rejecting the relevance of splenic hypertrophy, we retained the idea that natural selection was involved in the qualitative difference between peak and declining populations. Third, we rejected the idea that had occupied my thoughts since prewar days that animals in the decline phase are nonviable, a rejection that spelled the death of the maternal stress hypothesis. Instead, fourth, we proposed that declining populations differed primarily in behavior from those in previous phases and only secondarily in other attributes.

These ideas took time to grow from Janet's original analogy between the persistence of sickle-cell anemia in humans and the persistence of an apparently hereditary blood disease in voles. Sickle-cell anemia is preserved because the heterozygotes have an advantage where malaria is rampant. Janet thought the deleterious condition her supervisor was trying to find might also be preserved through heterozygote advantage under adverse conditions among peak populations. In time, a revised solution became known as the polymorphic-behavior hypothesis and resuscitated the belief that Charles Elton had voiced many years before: "Another important result," he wrote, "of the periodic fluctuations which occur in the numbers of animals is that the nature and degree of severity of natural selection are periodic and constantly varying."[31]

I endeavored to cover my tracks by pointing out the supposed relevance of this heritable condition was no more than a vague speculation, but that some attempt had to be made to integrate this finding with general biological knowledge, especially as insects showed changes in viability that were evidently related to the regulation of their numbers.[32] Finally, I thanked my colleagues. The decision for me to be sole author was probably wrong. I could not have given this paper without Helen's field results, Janet's knowledge of population genetics and hematology, or Ellen Phipps's meticulous work in the animal room. My own feeling was that I should be the only one to stick my neck out. Of course, had this paper been the ecological equivalent of the Watson-Crick paper,[33] I would have felt bad about grabbing all the glory; but this was a risk my colleagues were prepared to take. As well, Janet had established priority through a note in *Nature*. She continued to study spleens, though switching to bank voles.[34]

Although wrong, this hypothesis gave rise to a thought that has continued to be fruitful, namely, that cyclic declines occur if and only if preceded by natural selection. Whether true of false is still unknown; but this explanation satisfied Whitehead's claim that ". . . almost any idea which jogs you out of your current abstractions may be better than nothing."[35] Furthermore, whether right or wrong, it had the advantage over previous ideas of being testable against a sophisticated body of theoretical knowledge. It could yet turn out to be ". . . a fruitful error . . . full of seeds, bursting with its own corrections,"[36] though it's taking its time.

The way Janet and I blundered into this polymorphic-behavior hypothesis should be contrasted with the elegant argument in its favor that, unbeknown to us, Voipio had advanced 7 years earlier.[37] Apparently there was no Finnish equivalent of the magisterial J.B.S. Haldane to persuade him of the foolishness of his ideas. Also, he was then unaware of Selye's General Adaptation Syndrome, which bemused me by offering so obvious an explanation for the observed qualitative changes among cyclic populations. Both influences account for the slowness of my own conversion to believing in the relevance of natural selection. The moral from this is that, to avoid the dangers of going to meetings and reading the literature, one should seek authority and distrust it.[38]

8.6 Fleas

What should you do when you find you have made a mistake like that? Some people never admit that they are wrong and continue to find new, and often mutually inconsistent, arguments to support their case . . . It seems to me much better and less confusing if you admit in print that you were wrong.

Hawking[39]

The most considerable difference I note among men is not in their readiness to fall into error, but in their readiness to acknowledge these inevitable lapses.

Huxley[40]

To understand why our lab voles developed large spleens, we must trace the history of the way we kept them. The prewar stock was looked after by Richard Ranson.[41] To avoid seasonal changes in food quality, he provided a standard diet of grain soaked in a vitamin mixture. The voles received water and hay *ad lib* and had their cages cleaned at regular intervals. George Leslie used Richard's records to construct the life table already discussed. Richard died during the war and so did his animals. After the war I built up a new stock of breeding animals and in another room introduced about 50 wild-born animals to study their survival. I gave them cabbage twice a week instead of the vitamin mixture, which someone had suggested might be toxic.

I needed to know when each animal died but had no time to search every cage every day nor to clean them regularly. Moreover, when their hay was removed these wild voles tended to leap out and kill or stun themselves on the floor. I therefore disturbed them as seldom as possible and merely noted each day whether they had displaced a wire laid across their oats trough.

To see if this rough-and-ready system was worse than the prewar system, I kept a few lab-born animals interspersed among the wild ones in cages that were treated the same way. To everyone's surprise, the lab-born animals lived longer than expected from the prewar results.[42] I therefore kept the new system going for the next few years and cleaned the cages only when they had been flooded by leaking water bottles. Later on, as leaking bottles remained a problem, I persuaded Denys Kempson to install a new watering system, after which the cages remained dry and seldom needed cleaning.

In my Cold Spring Harbor paper I had stated that "recent procedures differ, so far as is known, in one respect only, that is, that isolated pairs now include the descendants of animals which had been breeding under conditions of mutual strife."[43] I'm glad I put in the qualifier "so far as is known." For it had not struck me that improving the watering system was a change I needed to consider, and it was some time before I noticed a connection between voles with small spleens and clean cages and voles with large spleens and dirty cages.[44] The dirty cages, which were no longer damp, had become an ideal breeding ground for fleas. Adding DDT to dirty cages or transferring animals to clean cages showed that large spleens in the lab were mostly due to fleas. Later still, of two lab-born littermates released into the wild, I re-trapped one with a tick and a large spleen and the other with neither. We also worked with a parasitologist, who described three blood parasites that may also have affected spleen size, and there was probably yet more to the story than we discovered.[45]

If I had been smarter I might have spotted the connection between fleas and large spleens before Janet and I started testing the original spleen story. If I had, we would not have gone to Lake Vyrnwy and developed our next hypothesis as fast as we did. It sometimes pays not to be too smart.[46]

It may be good for a person's ego to present his research career as a well-designed logical progression, the product of a tidy mind. Pasteur certainly did.[47] Even Claude Bernard, who made "the wisest judgments on scientific method ever made by a working scientist,"[48] dressed up a first-rate inductive leap as a valid syllogism.[49] In describing his discovery of the effects of fasting, he committed the fallacy of affirming the consequent, that is, inferring the truth of the premises from the truth of the conclusion.[50] Not that it matters, especially as the inference, though invalid, happened to be right. Scientists, including Darwin,[51] often misrepresent their thought processes or conceal them. Said von Helmholtz "I naturally said nothing about my mistake to the reader.[52] Most people follow this good example: it's the experimental consequences that count. Leaving out the agony and ecstacy of one's creative processes will not seem fraudulent to other research workers. But from those who have done no research themselves it does indeed conceal "the state of imaginative muddled suspense which precedes successful inductive general-ization."[53]

8.7 The End Of Shock Disease

One reviewer thought that we couldn't have been very clever be-cause we went on so many false trails, but that is the way discover-ies are usually made. Most attempts fail not becuase of lack of brains but because the investigator gets stuck in a cul-de-sac or gives up too soon

Crick[54]

Before Cold Spring Harbor our last hope of cracking the cyclic code was that voles in declining populations were suffering from shock disease. So, besides weighing various organs, we measured the glycogen content of the livers of live-trapped voles. To reduce the chances of the animals suffering from the effects of capture, we kept going round the traplines throughout the day, killed the animals immediately, and put the livers into hot caustic. During these studies our populations were still out of phase, and one unexpected difficulty cropped up. Whereas animals in a peak population were active throughout the day, those in an increasing population were not. Thus, ani-mals from the increase phase entered the traps mostly at night and so spent longer in confinement than the majority of peak-phase animals. This differ-ence in activity is one I'd like to have followed up.[55] It might have introduced a spurious difference between populations and encouraged us to continue this line of work. Or it might have concealed a real difference. Luckily, it did neither: results were uniformly contrary to hypothesis—in violation of Mur-phy's first law that if anything can go wrong, it will go wrong.

Not until it became obvious that voles did not have shock disease did I reexamine the claim that snowshoe hares did have it. An earlier critique would have saved hundreds of man-hours. If I had been running the graduate seminar I ran during my teaching career, my students would no doubt have spotted the discrepancy in the original accounts. My trouble was that the pathological details were published first and the field evidence later. By that time many of us had become imprinted with the idea that shock disease was the answer to our problems. In defense of my own conditioning, I can only plead that no one else had spotted the inconsistency between the pathological and field evidence. My "Note on Shock Disease"[56] took a number of promi-nent workers by surprise, and produced a condition analogous to shock dis-ease among one of its authors, to whom I'd sent a first draft of the manuscript. Far from being overjoyed at seeing "another step in scientific insight,"[57] he complained (correctly) that I had picked on evidence against shock disease and dismissed the massive evidence in its favor. He considered this an unfair form of discrimination. Even one of my own colleagues accused me of putting back the clock.

Once the scales had fallen from my eyes, it was clear that the captive hares dying from shock disease in the spring were not a sample of those whose accelerating death-rates were responsible for the decline. The animals that had been crowded into the small room at the University of Minnesota were adults and yearlings, which, according to field evidence, had normal survival rates in the wild; but the animals chiefly responsible for the decline were juveniles. Furthermore, their high death rates had been confined to the previous summer, long before their survivors were brought in from the field. The only evidence that juveniles suffered from shock disease was obtained from a few in an outdoor enclosure. But, as with animals in the crowded room, the condition could be recognized only when the animals were at death's door. I therefore concluded that shock disease was a nonspecific terminal symptom and that the high death rates in captivity were an artifact. It may be possible to keep lab-bred hares under crowded conditions, but it's no way to treat wild ones. Their dramatic death rates nevertheless suggest that snowshoe hares, at least in springtime, have a form of mutual intolerance likely to be expressed in nature by threats, avoidance, and subtle but unknown forms of behavior that affect survival, reproduction, and dispersal. Years later Lloyd Keith and his colleagues found that reproductive rates began to drop off even at low levels of crowding.

The high hopes so many of us had had for shock disease illustrate once again the danger of getting excited about evidence that merely confirms one's preconceived notions and is not based on sound comparative evidence. The authors had overlooked the need for controls.

8.8 The End Of Nonviability

In the process of exclusion are laid the foundations of true induction, which however is not completed till it arrives at an affirmative.

Bacon[58]

Janet Dawson, as she then was, came to me for tutorials in her first year at Oxford and in 1957 for her D.Phil. At first she worked with me; she then worked with and later married Robin Newson, another student at the Bureau. I had the honor of proposing a toast to the bride, and in consoling the parents for the loss of a daughter, wish I had said they were gaining not one but two new sons.

On one memorable trip to Lake Vyrnwy we were accompanied by Frank Pitelka and Monte Lloyd, which (almost) justifies me in further dispelling the fallacy that we were a bunch of drab characters. Impres-

sed with the efficiency of sheep in keeping down the grass (see Figure 3.1), Monte took home two lambs to save him the trouble of mowing his lawn. Unfortunately, the lambs caused him trouble of a different kind, because they kept breaking into the neighbors' properties and wreaking havoc among the flower beds. As his long-suffering wife had said on the arrival of the lambs: "that's all we need, Monte." She already had to put up with his absent-mindedness, as did Marie Gibbs, our secretary. For when using the telephone in the coffee room, Monte used to turn off the noisy refrigerator and forget to turn it back on. After driving Marie to distraction by ruining our supplies of milk and butter he learned to drape a tea towel round his shoulders to remind him to turn the refrigerator on again. This scheme worked well except when lost in thoughts about crowding among invertebrates he meandered back to his office wearing the towel.

In September 1956 Janet and I expected that a hereditary form of hemolytic anemia would be the reason why voles were (as I thought) nonviable during a decline.[59] We had two populations out of phase with each other (see Figure 8.5), 'The Old Road' at a peak, 'The Forestry Commission' increasing. So we planned to make comparative observations on physiological condition. Alas, by 1957 the contrast had disappeared: instead of declining, voles on the Old Road remained at a peak for a second year, with high body weights;[60] voles on the Forestry Commission reached a peak, and in 1958 both populations declined in synchrony. Voles had had a 5-year cycle on the Old Road and a 3-year cycle on the Forestry Commission. (Much later, I suggested that the mildness of the winter of 1956–1957 might explain why the population on the Old Road population upset our experimental design by failing to decline in 1957. It might also have enabled the population on the Forestry Commission to catch up.)

In both winters 1956–1958, weights on the two areas were much the same in September and December (19–21 g), but survival of the September cohorts was worse on the Old Road than on the Forestry Commission, probably because the population was further advanced in its cycle. Predation, however, was perhaps more severe on the Old Road, where cover was less dense than on the Forestry Commission. Field work seldom gives unequivocal results.

With the strong contrast between peaks in 1957 and declines in 1958 we expected to find some associated differences in hematology, but nothing turned up. None of the observations made in December 1957 enabled us to predict the onset of the decline; and as the underweight voles seemed to be in good condition and survived well in the lab, we rejected the idea that they were nonviable. We therefore stopped trying to explain a decline solely in terms of the *delayed* effects of behavior during the peak and suggested that "some unknown aspect of their own behavior" had also had *direct* effects during the decline.

Although viable, the animals reproduced less well in 1958 than in the peak year: none of 11 females was pregnant in May 1958 compared with 38 of

54 that were pregnant or had littered in May 1957. We also had trouble getting females to breed in captivity in 1958. Unfortunately, we gave no details, and it was exactly 30 years before two authors showed that females from a declining population reacted differently to space constraints in the lab from the way their females reacted 1 year later (see Figure 12.3).[61] The authors found this surprising, as both groups of voles had been trapped from low-density populations. Despite my earlier brainwashing of one of these authors, he apparently still thought population density should be useful for predicting the behavior of the animals.

To return to our own animals, those we kept in the lab turned out to be a mixture; some grew as well in 1958 (decline) as had those in 1957 (peak); others put on hardly any weight however long they lived. The latter resembled those left in the field in 1958, the former resembled those present during the peak in 1957. Mean body weights in the lab were therefore lower than they had been the previous year. Although some animals reached maximum size in the lab, none did so in the field, which suggests that growth was inhibited. But failure to grow during several weeks in captivity seems too persistent a condition to be attributed to inhibition alone, and the poor reproductive performance also suggests that some but not all these animals differed in a more fundamental way from those present in earlier years. As late as May 1958, however, our samples included animals that made good growth in the lab. Perhaps these were 'increase' phenotypes that had survived the decline, or perhaps they were from submarginal habitats. Francis Evans,[62] the first Rhodes Scholar at the Bureau, had suggested in 1936 that such habitats might provide immigrants to start the next recovery.

Apart from the contrast in growth between field and lab, our data were much as before; but further descriptive data were not what we were after. We had expected to find an abnormal physiological condition, and as we had not, our main consolation was in having set a new speed record in refuting a hypothesis in population ecology. Also, after a search that had lasted from 1946 to 1958, we finally gave up looking for a physiological clue to the secret of the decline. Our evidence was admittedly weak, so the decision to abandon this line of enquiry was another judgment call.[63] Any changes in physiological condition, we decided, were an effect of changes in behavior and a symptom, but not necessarily the cause of a decline.

On the positive side, we had discovered that although some animals are inherently unable to reach maximum size during the decline, others are present but evidently inhibited until removed from the field. As changes in body weight were not the selective effects we had set out to study, we were lucky to have collected as many data as we did. Better data became available a few years later[64] from Berkeley, California, where breeding normally begins when the rains start in October and continues until they stop. (But in years of increase breeding goes on throughout the dry summer months.) Charley Krebs took a sample of decline-phase voles into the lab on 20–24 January 1964 and weighed them every 2 weeks at the same as he weighed those in the field. Ten weeks later, while the field animals had made poor growth compared with that in the previous year, voles in the lab were a mixture much as ours had been in 1958: some had become heavy, others had not (Figure 8.7).

The next chapter describes an event that signaled the passing of the torch from one generation to the next.

8.9 A Visit To U.B.C

It is a favourite popular delusion that the scientific inquirer is under a sort of moral obligation to abstain from going beyond that gener-alization of observed facts which is absurdly called "Baconian" in-duction.

Huxley[65]

In March 1959 I paid a visit that changed my life and involved me with someone I was to work with for the rest of my career—to its everlasting benefit. Charley Krebs (Figure 8.8) was then finishing his M.Sc. at the University of British Columbia and was about to start a Ph.D. on cycles

Figure 8.7. Distribution (%) of body weights of male California voles in the field during a peak in 1963 (juveniles omitted), a decline in 1964, and in the lab in 1964. From Krebs (1966: Tables 14, 15). Growth in the lab was not studied in the peak year; it would presumably have been as good as that in the field—as it was in the peak year studied by Newson and Chitty (1962: Figure 1).

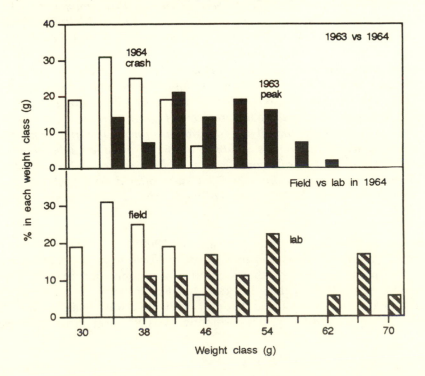

Figure 8.8. Longworth trap with one of its inventors (L) and one of its main users, Charley Krebs (R). Photo by Grant Singleton after the Sixth International Theriological Conference at Sydney in 1993.

> *in lemmings. The head of the Department of Zoology, Dr. Ian McTaggart Cowan, invited me to come over from Oxford to help set up Charley's program. The 3-fold consequences of this were that Charley went off to Baker Lake in the Northwest Territories that summer, that he came to Oxford in the winter of 1960–1961, and that I left Oxford in July 1961 for a career at UBC. I was anxious to teach, and thought life might offer higher things than spending the rest of it trapping voles. I remained in that exalted state of mind for 11 years before returning to research on a scale I now see was too limited to be of much use. Dropouts from research soon discover the truth of* sic transit gloria mundi.*
>
> *On this visit I gave three lectures, one of which sent Dr. Cowan into a profound slumber. In replying to his thanks for my inspiring remarks, I complimented him on his ability to absorb information subliminally. Had I known I was being considered as a future colleague, I might have been more circumspect. Luckily, Dr. Cowan offered me the job anyway and dispelled the belief of his colleagues that I would never again darken the doors of UBC.*

The gist of these lectures was published in 1960[66] and became a Citation Classic.[67] I pointed out once again that conventional wisdom about the regulation of numbers rested on the assumption that population changes were a function of density and that this gave "inadequate grounds for prediction, unless it could also be assumed that every individual was identical with every other." If, as seemed likely, this assumption was false, the theory of density dependence was equivalent to "counting a number of coins without observing their denominations or realizing that currencies may depreciate."

Population problems, I suggested, are of two main kinds that must be solved in different ways, but both by comparative methods. We need to know "(1) why population density does not go on rising indefinitely, and (2) why it varies from one type of environment to another." Instead of the original text figure, I will use a new one. The upper series of points in Figure 8.9 shows the spring densities of an imaginary population living in a good habitat; and the lower series shows the same for a poor habitat. One way of trying to understand these events would be to fit models to the data. As an infinite number of such models can be constructed, the only reason to prefer one to another is to see which makes the most successful predictions. Any that fail the test would be of purely local interest, if that.

An alternative to waiting 10 years to obtain one's first testable hypothesis would be to ask an experimental question, such as what is the difference between the increasing population A and the stationary populations a and B? Later on one could also compare c and d with C and D. For preference, one would not compare populations living at such different densities as those in Figure 8.9, because absolute densities that are similar in two different habitats (e.g., in year 8) are not comparable except in relation to their environments, past and present. In fact, however, we can often neglect differences in absolute numbers and make valid comparisons between population phases. At Lake Vyrnwy and Wytham, for example, phases were comparable even though peak densities were far higher at Lake Vyrnwy than in Wytham. In principle, comparisons between contemporary populations (A with a, etc.) are more reliable than successive comparisons within populations (A with B,

Figure 8.9. Numbers of two imaginary populations in good and poor habitats: respectively A, B... and a, b....

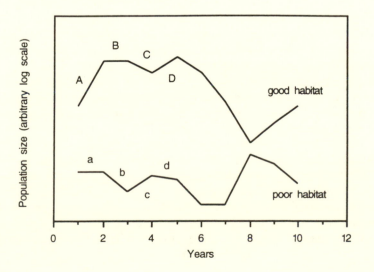

etc.) because of the confounding effects of weather and other variables. In all studies, there are are pros and cons for whatever one does, so it's best to have several irons in the fire.

It was, I thought, heuristically justifiable to assume that the enquiry into trends concerns an intrinsic property common to all species, namely hostility, which is the mechanism of threat and avoidance through which animals space themselves out.[68] Even if false, the assumption is testable, provided one has enough instances of a recognizable class of events. This is where cyclic populations are useful. A cyclic decline can be regarded as an exaggerated consequence of the phenomenon 'failure to increase.' Because it is recognizable, one can replace unfalsifiable propositions about *some* declines by falsfiable propositions about *all declines of a certain kind*. Expressed in experimental terms, the problem is to discover differences between populations that are increasing (the controls) and those that have stopped doing so (the phenomenon to be explained). Defining the problem in experimental terms might, I hoped, substitute an empirical for a rationalistic approach and remove verbal ambiguities over 'regulation,' 'control,' 'limitation,' 'determination,' 'drive,' 'density dependence,' and so on. Each of these terms can be replaced by an objective experimental statement.

Studying extrinsic factors seemed unlikely to help us generalize about the difference between increasing and other population trends. So, instead of studying mortality factors that differ from place to place and species to species, we should, I maintained, study behavior and see how it affects the probability of survival. Terminal fates would then be of purely local interest. The process I had in mind was analogous to senescence. Old animals differ from young ones in complex ways we need not quantify but whose effects we accept. The main effect is an increase with age in the probability of dying, even when old and young are living under the same conditions. I argued that individuals born in crowded populations differ congenitally from those born in increasing populations, and that the effects of extrinsic variables such as weather become more severe as numbers rise and quality falls.

There's probably something to the idea of qualitative change, but not to the idea that it's pathological. Indeed, if facts spoke for themselves, I would have given a different emphasis to my talks, for Janet and I had finished our field work, which, according to our 1962 paper, had convinced us there was nothing seriously wrong with the voles during their decline in 1958. The observations were there all along but their significance had not sunk in. Also, Ellen Phipps and I had yet to begin testing the idea that the essential difference between increasing and decreasing populations was in behavior and only secondarily in physiology. Our thoughts were thus in a transitional stage, but a change of emphasis was evident in a review[69] of some work on tsetse flies. Current views had failed to suggest "causes or mechanisms which prevent an indefinite increase of these insects when external conditions are favourable," and it seemed possible that tsetse flies, in spite of their low population density, had a form of behavior that regulates their population density through dispersal.

If ideas worked out on voles could be tested on insects, we could now dispense with mechanisms that were purely mammalian and look at more general mechanisms, such as dispersal, dispersion, and natural selection. There was no need to look for a density-dependent increase in the severity of any mortality factor. Animals forced into marginal habitats, or harassed by conspecifics within their own habitat, would die from starvation and exposure to predators, disease, poor food, and bad weather. A change in behavior rather than physiology alone made it easier to see how animals would become more susceptible to predation.

One restriction on the generality of these claims was that they apply only to closed populations, as "will become obvious," I said. I doubt if it did, and it was some time before I mentioned the problem again. Clearly, I thought, self-regulation of numbers could not be tested on cyclic populations subject to invasion from habitats whose populations were noncyclic or in a different phase.

Besides enquiring about trends or phases, we may also wish to know why populations differ in average density. We shall then be concerned with many factors, and the experimental question is different. We shall need to know which differences between good and bad habitats are associated with differences in abundance. Common sense shows that such differences may be explained in countless ways: by differences in numbers of predators and parasites of a variety of species; by differences in the incidence of disease; by differences in depth of soil, pH of water, concentration of minerals; or by differences in vegetation, moisture, agricultural or fishery practices, and so on. One or several of these variables might be relevant to differences in abundance *between* habitat types and to *irregular* differences within them. But none is necessarily relevant to the study of *recurrent* differences *within* populations. Unfortunately, as all populations are affected by both extrinsic and intrinsic variables, it's seldom easy to separate their effects.

To assume that all species are *capable* of regulating their own numbers is entirely different, I warned, from believing that all populations are in fact so regulated. In particular, the assumption that a self-regulatory mechanism has been evolved by natural selection implies that it has been adapted in relation to a more or less limited range of environmental conditions. In unnatural or atypical situations, therefore, the mechanisms will not necessarily prevent abnormal rates of increase or recurrent food crises. I had in mind small captive populations and those affected by agriculture or forestry; I did not foresee the 'Krebs fence effect,' in which, even in enclosures as large as 0.8 ha, the animals failed to regulate their numbers and continued to increase until they starved to death. In testing these ideas, therefore, a *sine qua non* is to have evidence of the kind given in Section 5.8 that one is in fact dealing with a population or phase to which the ideas apply. Once the general pattern of a cycle was understood, the effects of accidental variables, thas is, those that merely modify a cycle, could be sorted out. Quantity and quality of food were high on the list, and a chance to study one of them was soon to come.

I wound up as follows: "Voles probably exemplify a general law that all species are *capable* of limiting their own population densities without either destroying the food resources to which they are adapted, or depending upon enemies or climatic accidents to prevent them from doing so" (emphasis added).

At about the same time Wynne-Edwards,[70] impressed by the general absence of 'overfishing' among natural populations, advanced a similar concept of self-regulation but claimed it was achieved through group selection. I myself welcomed the massive evidence that animals can limit their own abundance through social behavior but considered it wrong to explain the phenomenon in terms of group selection. Charles Elton and David Lack, however, mounted vigorous attacks on both concept and explanation. Charles thought Wynne-Edwards's inexperience of field conditions had led him to oversimplify the way in which animals live and to neglect the importance of predation, parasites, and pathogens: "The theory is set forth with enthusiasm, often pontifically (if a bishop can wear blinkers), sometimes in a sort of Messianic exaltation which admits of no other processes affecting population levels." [71]

David Lack expressed his indignation at frequent intervals throughout his book and in a 6400-word appendix, which wound up as follows:

> Most of the behaviour considered epideictic by Wynne-Edwards can be satisfactorily interpreted in other ways . . . there is no reason to suppose that reproductive rates have been evolved to balance mortality rates; reproductive rates are explicable through natural selection, and the balance between birth-rates and death-rates can be attributed to density-dependent mortality . . . there is no need to invoke group-selection.[72]

It is unfortunate that Wynne-Edwards nailed his colors to the mast of group selection. The energy his critics[73] have expended on repudiating an untenable mechanism has diverted attention from his concept that "external checks" might indeed be "hopelessly undependable and fickle in their incidence."

9

From Wytham Woods to Baker Lake, 1959–1962

9.1 Behavior During The Decline

Knowledge, as opposed to fantasies of wish-fulfilment, is difficult to come by.

Russell [1]

Ellen Phipps looked after the voles used in the work described in Chapters 6 and 8, then helped with the field work described in the present chapter. Most research workers rely on their technicians, but I know of no one who feels as did Francis Bacon when he wrote:[2] ". . . I hold it to be somewhat beneath the dignity of an undertaking like mine that I should spend my own time in a matter which is open to almost every man's industry." Technicians soon learn to do a better job than their bosses, who rely on them to carry on when they have to tear themselves away from the lab or the field to do things they would prefer not to, such as writing research proposals or attending committees. I could not have asked for anyone more conscientious and hard-working than Ellen. She took a great interest in the work and kept experiments going while I was away or otherwise occupied. It was no more than her due to be coauthor of the papers resulting from our joint endeavors.

Like all previous field work, my last study at Oxford[3] was a test of conclusions from the study before. The idea that voles are nonviable during a cyclic decline had been the guiding light since before the war; but the preceding study with Janet had thrown doubt on that and suggested that, after all, the environment must also be where the trouble lay. The environmental factor that Ellen Phipps and I now set out to study was different, however, from the factors usually classed as environmental. This time it was the environment created by the behavior of the voles themselves.

Every individual, whatever its physical surroundings, lives in a unique environment.[4] Each animal is surrounded by a different set of neighbors. Thus, the environment of one sibling differs from another's, that of a subordinate from a dominant's, that of a male from a female's, and that of a sexually active animal from an immature's. So, if behavior affects survival, we must somehow measure these forms of behavior; prediction, the *sine qua non* for scientific respectability, is impossible from a knowledge of density alone, as we have already seen.

For the previous 2 decades I had supposed that declines took place even in a benign environment; but it now seemed that the field environment was less benign than we had inferred from its physical condition. The first thing to do, therefore, was to see if changes in survival could plausibly be related to changes in the social environment.

I was now more aware that, although some populations seemed to fade away gradually, others went down suddenly. It may seem surprising that I took so long to realize the significance of this, as the present retrospective account has emphasized the suddenness of a decline as one of the chief characteristics of a cycle. The most likely explanation for this mental block is that survival rates had been smoothed over several weeks, a consequence of our having been unable to monitor declines week by week. The problem is obviously different when animals disappear in a few days instead of a few weeks, and I should have realized that loss of viability, which would be consistent with a gradual decline, was unlikely to occur abruptly. We therefore hoped that we could do indirectly on voles what David Jenkins had done directly through observing his partridges, namely, show that changes in survival or emigration were associated with changes that were inconsistent with changes other than in behavior.

Now that samples of animals were no longer needed for autopsy, we could switch most of our work from Lake Vyrnwy to the smaller areas that Gillian Godfrey had found suitable on our back doorstep. In August 1959, therefore, Ellen Phipps and I began to follow changes in survival in much greater detail than had so far been possible. Instead of sampling at intervals of 4–8 weeks, we now sampled a population every week.

All small mammals had been scarce in Wytham in 1958,[5] but by August 1959 vole populations were nearing the end of a peak breeding season after a winter of reproductive activity. Survival during the next winter remained constant, less than 2 animals out of 10 being lost every month (a *minimum* survival rate of $P* = 0.92$ per 2 weeks). Then, towards the end of February, two cohorts of males (first the oldest and then the youngest) disappeared almost completely, and 4 weeks later the rest of the males lost almost half their numbers within 2 weeks (Figure 9.1). Meanwhile the females and both sexes of bank voles continued to survive at the same rate as before. Two months later, however, the female field voles lost 60% of their numbers in the 2 weeks after 25 May, and by June the population had reached the bottom of its cycle.

This decline was followed by rapid recovery in mid-July; but having been fooled in the past, I knew that recovery was only temporary, and that by next spring voles would be even scarcer. I was fooled once again, however: the decline did not continue, and body weights jumped back to their peak

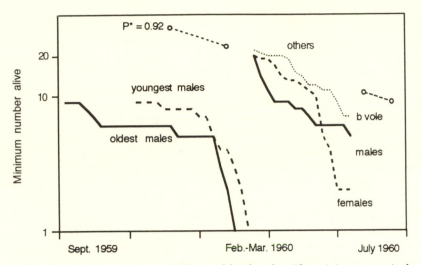

Figure 9.1. Declines in numbers of voles and bank voles. The minimum survival rate over winter was uniform at P* = 0.92. From Chitty and Phipps (1966).

value, which made this the first-recorded 2-year cycle.[6] Male body weights in April were as follows (those for peak years 1959 and 1961 are in bold type): 21, **34**, 23, **32** g. This immediate return to a peak did not, however, detract from the discovery that the decline of 1960 was associated with a series of losses that were too sudden to be explained by changes in physiology alone. Nor, by themselves, could environmental variables such as disease, predation, food shortage, or weather have singled out one cohort without affecting the others. Similar sudden losses occurred in 1961, except that the females declined before the males. It was no surprise to find these sudden losses confined to one sex at a time; but it was a surprise to find different age groups disappearing at different times. It suggests that antagonism in spring is most intense between animals of the same age and sexual condition.

As we expected some of these losses to be due to dispersal (in spite of previous evidence to the contrary), we removed all residents from an adjacent area of about 2 acres, thinking that potential dispersers might previously have been inhibited from entering occupied ground. To our surprise, most deaths seem to have been *in situ*; as only one of the missing individuals took advantage of this empty habitat, though another turned up in traps John Clarke had set about one-third of a mile away.

Individually, these changes in numbers could be explained in several different ways; but an explanation consistent with them all (and therefore the simplest) was that they were due to age- and sex-specific changes in behavior. Two objections to this solution are, first, that one can account for anything whatever, merely by assuming sufficient heterogeneity of interaction within and between different cohorts. This meant that the hypothesis was difficult to refute until we could devise ways of studying behavior in the field. Second, hostility seemed an implausible explanation for losses that affected small,

widely spaced, and well-established cohorts. Nevertheless, the results supported the view that progress would be impossible without critical studies of
behavior and genetics.

While supporting the idea that changes in behavior were associated with
changes in survival, the results gave negligible support to the idea that
changes in gene frequency were also involved. We had yet to find out whether
offspring recruited into a crowded population differed genetically from those
recruited at other times. But it so happened that 1959, which had been a
peak year at Wytham, had been a year of increase at Lake Vyrnwy. The two
sets of animals breeding in 1960 thus differed in the conditions under which
they had been born and raised, a difference associated with characteristic
differences in mean body weight.[7] In the spring of 1960 we therefore set up
breeding pairs both from Lake Vyrnwy and Wytham and recorded weaning
weights of their young. Those from Wytham pairs had a bimodal weight
distribution, many being underweight (in spite of having been normal at
birth), whereas those from Vyrnwy pairs were normal (Figure 9.2). On average, the Wytham females were 6 g lighter postpartum, but relatively more
had litters of six and seven, their average litter size being 4.6, whereas the
average for Lake Vyrnwy famales was only 3.5. None of this made much
sense, but may have had something to do with the rapid recovery from the
Wytham decline.

We now needed to know how offspring from the different phases would
survive in the wild; so in November 1960 we released 23 animals of each
kind on the 2-acre plot previously cleared of residents. We placed each litter
in a well-stocked underground cage, from which the young escaped through
a hole in the bottom and burrowed their way to the surface. About half of
each stock was lost at the start, and although the remainder had a good
survival rate (P* = 0.94 per 2 weeks), only four of each kind of female were

Figure 9.2. Weaning weights, in litters of four, of voles born in captivity to parents
from an increasing population at Lake Vyrnwy and a declining population at Wytham. From Chitty and Phipps (1966).

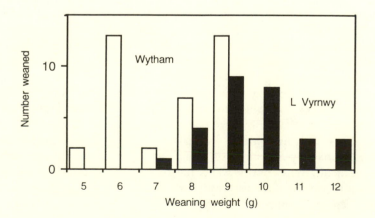

left by March 1. Three of these were pregnant, all descended from parents born in the expanding population at Lake Vyrnwy; but through lack of time we discontinued these observations and removed the animals.

These preliminary observations were faulty in several respects and provided no evidence that differences in body weight of the offspring were due to genetic rather than maternal or geographical differences. But as it was likely to be some time before I could do proper replicates, I thought the results worth describing. Little did I know it would be 12 years before I made a similar attempt to find out if animals born in an increasing population differed genetically from those born in a stationary or declining population. It, too, was unsuccessful.

Besides monitoring the population of field voles, we did the same for bank voles, whose adults suffered no comparable losses in spring, and changed in numbers more gradually throughout the year. Their young males, however, resembled young field voles in suffering high losses during both breeding seasons, presumably owing to attacks by adults. Figure 9.3 shows the numbers released in the 1960 season and the relatively small numbers still alive 4 and 8 weeks later. After September 1959 and October 1960, however, about 70% survived the first 4 weeks, coincident with the end of breeding and associated hostility.

Neither on this occasion nor in 1948–1950 did I look at the bank vole results as carefully as I should have. The two species are alike in some of their behavior, but I should have realized that they probably differed in ways that might have explained why one had had a decline and the other had not. The populations shared the same physical environment, so it might have been easier to find relevant differences between species than it had been to find relevant differences between cyclic populations in Scotland and the noncyclic population at Corris. So these were lost opportunities to overcome the difficulty that the greater the difference in conditions, the harder it is to separate relevant from irrelevant variables. However, one can't do everything, and with the change of job in 1961, I was hard pressed to do as much analysis as I did.

In spite of having had no low phase, this 2-year fluctuation was presumably an intermediate variant between a 'typical' cyclic decline and a mere seasonal one. The low body weights in 1960 were typical of a cyclic decline, and the bimodal weight distribution of the Wytham progeny suggests that their parents were a mixture: some perhaps typical of the increase phase, some of the decrease. The absence of a low phase might therefore have been because too many increase phenotypes had survived the winter and so enabled the population to have a 2-year instead of a typical 4-year cycle. All was not lost, however, as explained in the next section.

To sum up: it seemed clear that behavior played a direct role during the decline itself and not merely a delayed role due to its effects on physiology during the peak. It also seemed that hostility might be confined to some cohorts without affecting the survival of others.

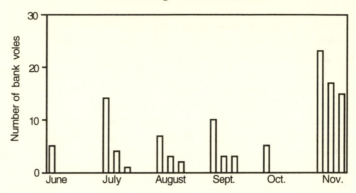

Figure 9.3. Number of young male bank voles released and number alive after 4 and 8 weeks. Plotted arithmetically because of the two zero catches. Data pooled for November 1960–January 1961. From Chitty and Phipps (1966).

9.2 Food, Weather, and Body Weight

Rather than express extravagances, [it is] better to pronounce that wise, ingenuous and modest sentence, "I know it not."

Galileo[8]

In trying to understand differences between phases we had been too busy to study changes in accidental variables, that is, those that modify a cycle but are not necessary for its occurrence. Food and weather are two of the most obvious modifying conditions, and thanks to the help of two visitors in 1960–1961 I was able to study the relation between a vole population and its food supply. Also, thanks to a freak winter I was given a hint about the effects of weather. One of our visitors, as already mentioned, was Charley Krebs. The other was David Pimentel,[9] one among the few population ecologists besides Charles Birch[10] who at that time believed that natural selection plays a part in the regulation of animal numbers. Six years later we produced an article entitled "Food supply of overwintered voles," for which David had done most of the work.[11] Our immediate aim was to determine the area of grassland required to support a vole at a time of year when food is scarcest and a decline in population density was expected. A second aim was to see how body weights responded to surplus food.

The most direct way of achieving the first aim was to enclose a vole within a fenced plot and see how long it took to use up its food supply.[12] We had hoped that a slight loss of body weight would tell us when food was running out and we should remove the voles. Unfortunately, the first three lived without weight loss and died suddenly in the next 24 hours. Thereafter, we removed the voles as soon as they had reduced the vegetation to the point at which it seemed unlikely to keep them alive any longer. The time to starvation was therefore underestimated; but the early deaths showed how sensitive these animals might be to unfavorable conditions. We concluded that no vole need starve if 2.1 sq ft of herbage were available each day. With only 50 animals per acre (including bank voles), they were destroying only 0.24% per day of their above-ground food supply. This was in February and March 1961; and during the decline in April plenty of new vegetation had already sprouted.

These observations confirmed conclusions from less elaborate studies, namely, that starvation is not a necessary condition for a cyclic decline. But the size of the surplus—a whole year's supply of green vegetation—surprised even me. We did not suggest that supply would always outstrip demand to such an extent, and knew that many more studies of this sort would be required. But the results also gave us some ideas about the unexpected 2-year cycle on Rough Common.

One possible explanation was that survival (at least among certain phenotypes) had been better than usual owing to the mildness of the winter of 1960–1961. As much as 20–40% of the vegetation was still green in February-March, which led us to ask what would have happened if, in the normal course of events, most of it had been killed by frost. With a smaller surplus of food, hostility might have been more intense; but we refrained from guessing whether fewer animals would have come through the winter or whether the spring decline would have been earlier or more severe. The important point was the realization that the effects of 'crowding' would vary according to the state of the extrinsic variables (see below).

Our second aim was to see whether supplying high quality food would prevent body weights from remaining at the low level expected during a decline (I predicted it would not).[13] For comparison with the main study area, we chose a small, fairly isolated strip of grassland similar to that elsewhere on Rough Common, and from February onwards supplied the animals with a surplus of oats and carrots. By bad luck there were never more than seven *Microtus* males and one female, but by good luck they were out of phase with those on the rest of the area. In fact, body weights behaved the way I had expected all of them to behave: they were low in March and, despite the extra food, remained as low in April as those in the decline of 1960.[14] One of these individuals had a feeding station all to himself and visited none of the others in the 22 times he was captured. In spite of being isolated and having plenty of food, he nevertheless failed to exceed 21.5 g in the field or 23.5 g during a further 2 months in the laboratory. Elsewhere on Rough Common all males exceeded 28.5 g after the end of April.

This pilot test might well have gone the 'wrong' way, thanks to the unexpected recovery in numbers and body weight on the main area. If we had given extra food to those animals, they might have put on even more weight than they did. And the isolated, unfed, population would have given the false impression that their low body weight could indeed have been improved through good feeding. This was a lucky escape, which in theory we did not deserve. In practice it's so difficult to replicate population experiments that one must make the best of one's bad jobs. As here. Though based on a small sample, these results confirmed all previous evidence that body weight is low during cyclic declines and may stay low even on a high-quality diet. The refractory nature of a typical low phase is the main reason for doubting whether it can be accounted for by any process except natural selection. Later work has confirmed that one cannot stop a decline by providing surplus food.[15]

Another undeserved piece of luck was the mildness of the winter, which suggested new ideas about the effects of weather. At one time I thought a decline should continue however benign the physical environment. This now seemed less certain; the exceptionally mild winter might have prevented the decline expected in 1961—and the mild winter of 1956–1957 might also have prevented the decline on the Old Road at Lake Vyrnwy. The effect of a mild winter would be to change a poor to a good habitat (see Figure 8.9), which would mean that population density was now lower than usual in relation to its food and other extrinsic variables; hence, that survival would be better than usual. This effect might be due solely to the milder weather; but if due to a better food supply, it contradicted other evidence that one cannot prevent a decline by adding extra food. Regretfully, therefore, we had to admit how little we understood what was going on. The study ended not with a bang but a whimper.

9.3 Experimental Predation and Other Tests

A theory is a good theory if it satisfies two requirements: It must accurately describe a large class of observations on the basis of a model that contains only a few arbitrary elements, and it must make definite predictions about the results of future observations.

Hawking[16]

The Australian student, M.E.B. "Mike" Smyth, was a Rhodes scholar from Adelaide, where the zoology department, under the chairmanship of H.G. "Andy" Andrewartha, taught with heavy emphasis on methodology and statistics. Mike was initially attracted to the Bureau by the prospect of working with Chitty, whose hypothesis about the mechanism of vole cycles intrigued him. But by the time he got to Oxford, Dennis was packing to go to British Columbia, so Mike was obliged to work with Mick Southern. He set out, nevertheless, to undertake a project suitable for one of Chitty's students. . . . [17]

In 1960 I made certain predictions[18] that, if falsified, would throw doubt on the idea that populations are self-regulated through changes in physiological condition. The predictions are summarized in Figure 9.4.

The first person to accept the challenge was Mike Smyth,[19] who tested prediction 2a that, if numbers are kept down by persistent cropping, a population will remain indefinitely in the increase phase. Mike's main work was on the bank vole, one population being the control for a cropped population. Figure 9.5 shows the numbers occurring naturally on his 1.8-ha control area

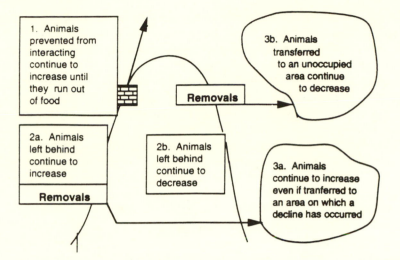

Figure 9.4. Predictions made to test the hypothesis that behavior before the decline has a delayed effect on physiological condition—but not on behavior. (Chitty, 1960).

Figure 9.5. A cropping experiment by Smyth (1968). For simplicity, only two pairs of arrows are included to show the monthly reductions and recoveries in numbers of bank voles on the experimental area.

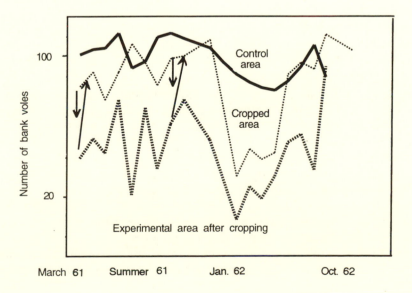

and those before and after monthly removals from a similar-sized experimental area. In spite of the removal of 485 bank voles between March 1961 and January 1962, numbers climbed back each month to what they had been before. Between January and June, however, fewer surplus animals seem to have been around, and the experimental population remained at about half the density of the control. Then, after breeding started, Mike was again unable to prevent the experimental population from being restocked by immigrants, at least up to a restricted level.

Heavy 'predation' thus had little permanent effect on population density, and the fact that numbers kept coming back to about the same level is consistent with the idea that animals can keep their numbers down by limiting recruitment from within and outside the area. But by bad luck this self-regulation had no delayed effects, for the 1962 decline on the control area was merely a prolonged overwinter decline (as far as one could tell) and not the sudden cyclical kind Mike had hoped to study. Mike also experimented with wood mice and was equally unsuccessful in keeping numbers down on the removal plot. But by good luck, his control population did suffer a severe decline followed by extreme scarcity until the end of the breeding season. This was the sort of decline and scarcity that, according to hypothesis, was preventable through heavy trapping, but if and only if the experimental plot had been a closed system. As it was continually flooded with animals from declining populations, the experiment was almost certain to fail. Nevertheless, Mike was right to point out that

> The decline in the numbers of wood mice on the experimental as well as on the control area between November 1961 and December 1962 is a more serious objection to Chitty's hypothesis, since it was almost certainly the kind of decline for which his hypothesis was proposed.

In his thesis[20] Mike doubted my hypothesis mainly because he doubted that "all declines (or most, calamities aside) belong to a single class of events." Unfortunately, I was not there to tell Mike he had got the idea backwards and overlooked the need to distinguish cyclical declines from all others, such as the inevitable declines during a nonbreeding season. He wound up as follows:

> Because of the immigration . . . it would be difficult to claim that any of my evidence refutes Chitty's hypothesis. If it can be refuted, it can only be done by a better controlled experiment. My results do show that it is difficult to do field experiments to test hypotheses about such complex phenomena as these oscillations. But it will not be impossible. [21]

A similar experiment was also unsuccessful, however. In 1963 Charley Krebs cropped a population of California voles every 2 weeks for a year, but in spite of removing the impressive number of 1758 animals from a 2-acre plot,[22] he, too, found it impossible to keep numbers down and so was unable to prevent them from reaching a peak and declining.

Of the remaining predictions in Figure 9.4, number 1 implies that if hostility is prevented (for example by putting a fence round each pair of voles) the population will no longer be self-regulated. The discovery that self-regulation breaks down among enclosed vole populations[23] is consistent with the idea that families keep to themselves, as shown in Figure 6.7, perhaps because dispersal is reduced or absent. Prediction 2b—that reducing population density will not stop a decline once it has started—has been tested on voles in Finland and on lemmings in Sweden, where workers removed females during a decline. Both declines continued anyway.[24] Number 3a predicts that plenty of food will be available for an increasing population transferred to an area on which a decline has occurred. This prediction turned out as expected, as did Number 3b, which states that populations will continue to decline however good their food supply: it was verified for a vole population provided with lab chow,[25] and next by the continued decline of defoliating insects on areas they had not defoliated.[26] The quality of the larvae had evidently been affected earlier in the cycle. A similar conclusion seems to fit the tent caterpillar, which resisted attempts to produce out-of-phase populations. Although introduced to areas with abundant food, "the introduced populations declined at the same time as their source populations." [27]

Charley Krebs later expanded the number of potential falsifiers,[28] and Watson & Moss laid down other criteria by which to judge the hypothesis of self-regulation.[29]

Mike's claim that cropping experiments are not impossible was eventually corroborated. Figure 9.6 shows that a decline in numbers of red grouse was delayed for 4 years by cropping.[30] The best evidence is for 1985, when num-

Figure 9.6. A cropping experiment by Moss and Watson (1990, 1991), showing the number of male red grouse before (●) and after (○) removals.

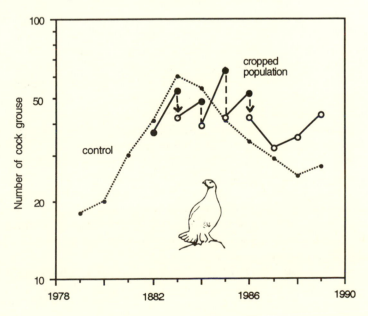

bers were almost identical on both areas and went up from 42 to 52 on the experimental area and down from 41 to 34 on the control area. Next year, however, both populations declined, presumably because the experimental area was flooded by immigrants from the declining population.

A neat experiment such as this is worth more than a thousand words, but until the experiments above have been replicated, perhaps more critically, they are unlikely to convince those who still deny that populations are capable of regulating their own numbers.

9.4 Lemmings at Baker Lake

To the same natural effects we must, as far as possible, assign the same causes.
Newton's second Rule of Reasoning in Philosophy[31]

Besides the ideas developed at Oxford, two rival schools of thought had developed, and the time was ripe for all three to be tested by an independent worker. Before starting his Ph.D. Charley Krebs had had his ear bent in a series of talks at UBC, not only by me but by Frank Pitelka. He was also familiar with the views of John Christian. His first purpose, therefore, "was to describe events of the lemming cycle of the Barren Grounds;" and his second "was to explain these events in a comprehensive theory" and find out "which of the current explanations [were] inadequate." And into four short summers Charley managed to cram both descriptive and experimental work similar to that undertaken by many people in various studies spread over many years.[32] His success in doing so earned him the Terrestrial Publication of the Year Award for 1964 from the Wildlife Society.

The vole studies had suggested which aspects might be omitted from and which included in the search for an explanation of the lemming cycle. Neither parasites nor infectious diseases seemed relevant; predators and movements might be; physiology, behavior, and genetics were each in the running; and food still had its backers. Working in an area barren of research facilities, Charley postponed work on other aspects of the problem until going to Berkeley on a Miller postgraduate fellowship.[33]

Two species of lemmings were present at Baker Lake (see Figure 1.5), the brown lemming in the wetter habitats and the varying lemming (see frontispiece) in the drier ones. As in the vole, peak populations suffered a spring decline, from which they returned to a high density similar to that of the previous winter (Figure 9.7). Again as in the vole, the next winter's population began its decline before the breeding season and continued it during the spring. Although breeding was resumed as usual with the melting of the snow, all populations declined. Some continued to decline through the summer, some recovered before the fall.

Body weights also changed during the cycle in much the same way they

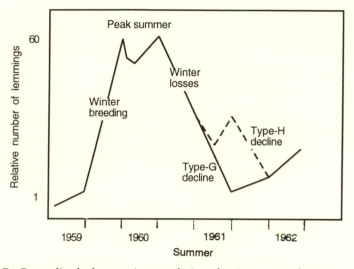

Figure 9.7. Generalized changes in population density among lemmings at Baker Lake, N.W.T. Type-G decline on main area; type-H decline on 'Seccond Island.' From Krebs (1964a).

had in vole populations (Figure 9.8). Large animals began to show up during the increase phase in August 1959, were present during the peak spring, and absent from the main study area during the decline in 1961. Some large animals were still present, however, on the area that recovered its numbers that fall, and a few of moderate size were present on the main area when numbers picked again up in 1962.

The main difference from the vole story was that winter breeding was more pronounced during the increase phase than it had been in voles; otherwise, the data to be explained were essentially the same, and an explanation in terms of changes in behavior was consistent with them. Being consistent with the data is, of course, a far cry from surviving a critical test, but is better than being inconsistent, which was the fate of the other two explanations. The evidence against them was as follows.

To study the food supply, Charley fenced off several areas, compared the standing crop inside with that outside, and estimated that lemmings had used up less than 17% of available supplies by the end of any of the growing seasons 1959–1962. To study the overwinter supplies, he ran transects through different habitats and estimated that no more than 30% had been used during 1960–1961. As well, he used a fat index to measure the condition of the animals, and at the time of the decline found it equal to or higher than that in other years. He therefore rejected the hypothesis that the decline in numbers of lemmings was due to food shortage; nor was there any evidence of malnutrition.

To detect changes in the physiological condition of the animals, Charley weighed over 7500 organs from about 3000 lemmings. Mere weights may

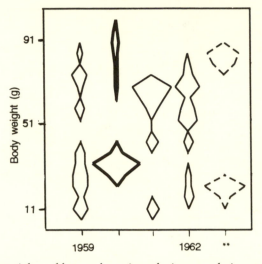

Figure 9.8. Body weights of brown lemmings during a cycle in numbers on the main area at Baker Lake, N.W.T. and during a Type-H decline (broken outlines) in 1961** on 'Second Island.' Thick outlines show peak year. From Krebs (1964a: Tables 40 and 44).

not, or course, reflect physiological condition; but a progressive change in weight would encourage one to look for associated changes in condition. No such encouragement came from data on adrenals or spleen. Testes, however, did change in weight during the cycle: they were heavy in adults and sub-adults of both species in the increase summer of 1959 and light in young males in the peak summer of 1960. During the decline of 1961 the young of both species again had light testes, although the animals were no more crowded than they had been during the start of the increase 2 years before.

> Because this inhibition of gonadal development in young male lemmings oc-curred in the summer of decline as well as in the peak summer, density per se cannot be the factor directly involved here, but rather the important variable must be capable of acting at very low densities in the decline.

Although there was no evidence that any age groups were in poor physio-logical condition, something was going on to depress the maturation of young males. A clue came from the amount of wounding:[34]

> In both species the peak summer of 1960 had a significantly higher amount (P <.01) of wounding in June, the time before the summer young have appeared. In spite of very large differences in density between years, the amount of wounding found in July and August was not markedly different. This implies that fighting may occur even at very low densities.

Also, during the increase in August 1959, Charley collected the only sam-ple in which some males of the year had matured and the rest had not. Six out of eight of the mature males were wounded compared with only 1 out of 11 immatures. Maturation of young males seems to have led to an increase in fighting. Failure to mature in other breeding seasons, Charley suggested, is perhaps an adaptation to prevent it.

Young females, unlike their brothers, were not totally inhibited from maturing during the peak and decline, but, as at Lake Vyrnwy, matured later in life during the peak than they had in the year before. All animals stopped breeding early in the peak year.

I shall not try to summarize the rest of the field data, except to note that predators were relatively scarce during the decline and seemed to have little effect on it, and that dispersal onto the lake ice occurred in the peak year only, being confined to the 9 days before June 4, 1960. No "migrations" took place in the other years; the decline phase, therefore, was not associated with the legendary suicide march.[35] The Discussion included a review of the relevant literature and ended as follows:

> One of the interesting points that has come from this lemming work is the similarity between this lemming cycle and the cycles in *Microtus agrestis.* . . . That similar types of events should occur in two such different ecological situations argues quite strongly for a unified view of cyclic processes.

In another paper[36] Charley presented an analysis of almost 2000 skulls of the two species. This was the closest he could come to testing the idea that qualitative changes associated with the cycle might have a genetic component. He found that skulls of animals caught during the decline differed from those caught during the peak:

> Lemmings of a given body size do not have the same size skull at different phases of the population cycle. Why should this be? . . . [One] explanation of these shifts attributes them to changing age structure. Thus an adult of a given

Figure 9.9. Cycle in skull size of lemmings during a cycle in numbers. Mean deviations from common regressions measure the size of the skull in relation to body weight. Two calculations were made each year: one for adults, one for young. From Krebs (1964b).

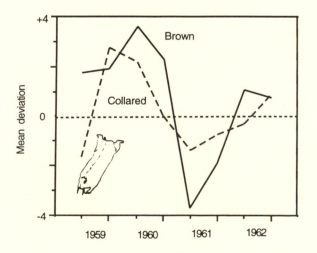

weight may be of quite different age in the different years of the cycle, and consequently if we could use age instead of weight, we might eliminate these shifts. The fact that summer young animals, known to be the same age, show these shifts rules out the possibility that differences in age structure are sufficient to explain them. . . . These skull-body changes in lemmings may be considered another index of a change in quality of the population with changes in abundance. . . . Lemmings of the same body size are not equivalent in the different years of the cycle. Why this is so and, in particular, whether these changes are genotypic or phenotypic remains unknown." [37]

Figure 9.9 shows how skull size (corrected for body size) had a cycle of its own in parallel with that in population density.

The good fortune of having become the supervisor of this workaholic student enabled me to turn my attention at UBC almost entirely to teaching and supervision, in the happy knowledge that he would soon solve the cycle problem. As he had not done so by 1970, in spite of the best efforts of himself and an outstanding group of students, I myself returned to the struggle, hoping to clear things up. I should have known better. [38]

Synchrony, 1924–1961

It is also a good rule not to put too much confidence in the obser-
vational results that are put forward until they are confirmed by
theory.

Eddington[1]

10.1 Lynx, Muskrats, and Game Birds

No studies of cyclic populations deal experimentally with the tendency for
weather to keep them in phase. There are good reasons for this tearing apart
of the seamless coat of learning. First is the practical difficulty of finding out-
of-phase populations within easy working distance of each other; at present
they crop up by luck only, and harassed field workers may be unprepared to
take advantage of them. Alternatively, when they do turn up, such populations may frustrate him or her by slipping unexpectedly back into phase. It
may some day be possible to manipulate populations into different phases,
wait in hopes of their coming back into line, and meanwhile guess which
variables to measure. The main difficulty, however, is theoretical. For while
there is no winter in the bounty of explanations for the biological changes,
there are no good explanations for their synchrony. I once suggested that
predators play a part in producing regional synchrony,[2] but there is no evidence that they are everywhere abundant enough to affect the length of a
cycle or change the behavior of the animals—as they may under some circumstances.[3] Weather therefore seems the agent most likely to explain the tendency for populations to go up and down at the same time.

Charles Elton had soon abandoned his hypothesis that biological cycles
are related to sunspot cycles,[4] though he left it to MacLulich to publish its
obituary.[5] On average, the sunspot cycle is about 1.6 years longer than the
lynx cycle (11.2 compared with 9.6 years). Two such cycles starting in phase
will therefore drift slowly apart, and when in phase in one area will be out
of phase in others. Reconciling irreconcilable cycles has challenged the ingenuity of a number of authors. MacLulich and Keith[6] give examples of these
exercises in futility. Keith remarks about one of them: "There is, in my opinion, no good reason why this theory should be considered anything more
than an example of how some data may be manipulated to produce nonsense
correlations." These are harsh words, but not harsh enough to have prevented a recent flogging of this dead horse. Nevertheless, weather distur-

bances associated with sunspsots may sometimes be associated with regional synchrony among cyclic populations.

In spite of examining all meteorological records he could lay his hands on, Charles Elton found no evidence of any cycle in weather. He may, as Peter Crowcroft states,[7] have continued to have faith that one existed; but if so, he kept quiet about it after the war. For if one assumes that the 10-year cycle is synchronized by a 10-year cycle in weather, one might as well well assume that the 4-year cycle is synchronized by a 4-year cycle in weather and that cycles among game birds, spruce budworms, larch budmoths, and other species are also synchronized by corresponding cycles in weather. A better alternative, I suggest, is to look for an explanation common to all cycles—in mammals, birds, and insects.

The lynx relies mainly on snowshoe hares for food, and its numbers are assumed to reflect changes in the numbers of its prey. Fur returns for lynx for northwestern Canada are shown in Figure 10.1.[8] Returns for the east are confined to the north shore of the St. Lawrence River, where the latest peaks in this particular series (1897–1939) were in 1925 and 1936, and cycles were more or less synchronous at both ends of the country. Around the turn of the present century, however, they were completely out of phase, and until 1916 bore no obvious relation to one another.

Other data for lynx go back to 1735 and suggest that peaks recurred with great regularity but took about 3 years to develop across Canada, those in the east often being later than those in the west. This sequence is contrary to later evidence for snowshoe hares but consistent with the idea that the cycles vary in length geographically and so will change from being later to being earlier in the east than in the west. Returns for lynx from Alaska did not follow the Canadian cycle closely.

Figure 10.1 also shows the 1897–1927 returns for muskrats for the northwest, the 1919–1938 returns for Alberta, and the same for Nova Scotia and New Brunswick combined.[9] The western returns show four periods of declining sales after peaks in 1899–1900, 1909, 1921–1922, and 1932 in Alberta and are similar in shape to those for lynx. Earlier returns from Alberta were complicated by the inclusion of skins from south of the boreal forest, where populations were out of phase with those in the north.

Data for Nova Scotia and New Brunswick include three periods of declining sales after peaks in 1919 or earlier, in 1924, and in 1930. A fourth peak may have been reached in 1938. These populations were out of phase with those in the west and may have had a shorter cycle; but given a series of only 20 years, we can't be sure.

Game bird cycles in the Maritime Provinces were about 3 years earlier than the 'main' cycle, and those in the extreme northwest and Alaska were about 3 years later;[10] and as shown below, the hare cycle had a similar east–west drift, which confirms the tendency for bird and mammal cycles to be in step. There was no evidence of a difference in length of the game bird cycle between Alaska and the Maritimes. In Europe, however, cycles varied in length from 3–4 years in Finland and Scandinavia, to about 10 years in Ice-

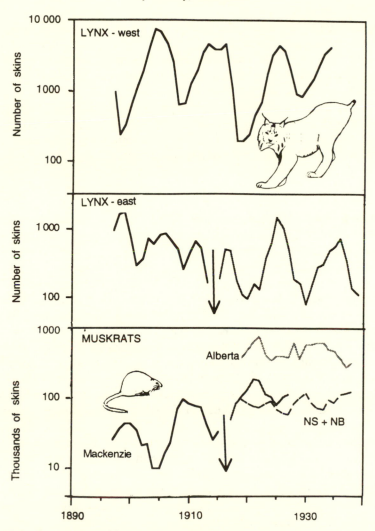

Figure 10.1. Fur returns of lynx and muskrats. Arrows indicate abnormally low returns. From Elton and Nicholson (1942a,b).

land. Cycles in the former U.S.S.R. and Spitzbergen resembled the short cycles of Britain and Scandinavia.[11] Among red grouse in northern England the average cycle length has been less than 5 years;[12] elsewhere it has been more commonly 6,[13] with a tendency to increase with latitude up to 8 years in northern Scotland.[14]

The '4-year' cycle in small mammals also varies in length, at least in Sweden.[15] It is said to be absent below 60° N and to increase from 3 to 5 years between 60° and 70° N.

So far, in struggling to understand the dynamics of their populations, vertebrate biologists have been too preoccupied to study a factor whose action they regard as density independent and therefore uninteresting. No credible theory can ignore the many exceptions, but neither can any credible theory ignore the tendency for independent populations to be in step. Moran tried, with doubtful success, to find a correlation between weather and the lynx cycle,[16] whereas Arditi[17] had better luck through introducing a larger array of weather variables.

It's now obvious that the synchrony problem is more complicated than it seemed at first and is unlikely to be solved without experimental studies of the way populations are affected by the most likely agent, weather. This is the only way to go if the 20-year run of data mentioned above is too short to be of much use.

10.2 The Snowshoe Rabbit Enquiry

Science consists of physics and stamp collecting.

Rutherford[18]

For the 17 years 1931–1948, the Bureau ran 'The Snowshoe Rabbit Enquiry,'[19] which was similar to the Canadian Arctic Wild Life Enquiry. Questionnaires were sent to trappers, Royal Canadian Mounted Police, postmanagers of the Hudson Bay Company, game wardens, National Parks staff, and others. Classically minded zoologists at Oxford thought much the same of this kind of work as Lord Rutherford would have thought of theirs. 'Mail-order zoology' hardly seemed like science; yet there was no other way of studying variations in the incidence of a "10-year" cycle over an entire continent.

The title of this enquiry was chosen to appeal to its correspondents, who usually called the animal a rabbit rather than the hare it is. Their mistake probably goes back to the early settlers, who, being poor taxonomists, based their classifications on trivial characters, for example, on the red breast of *Turdus migratorius,* known ever since as the American robin. The European robin is a bird of a different feather.

This enquiry enabled us to recognize large areas over which most hares were increasing at the same time (Figure 10.2).[20] But although increase was synchronous over large areas, it was not synchronous everywhere. On the contrary, hares reached a peak somewhere on the continent in almost every year from 1931 to 1946. The most extreme lag in our first cycle was between Nova Scotia, where hares declined in 1932–1933, and Alaska, where in places they declined as late as 1938–1939. In that year, according to Otto M. Geist (a man who recognized the value of comparative data), hares were scarce for 40 miles around Fairbanks. Whereas in the previous 2 years they had been abundant, only about one-eighth or one-tenth were now present. Where he had hunted dozens during the fall of 1937–1938, he now bagged only an occasional hare. The lag of Alaska behind the rest of the continent is confirmed by Philip,[21] also by MacLulich and Keith, whose analyses differ only slightly from the present one.

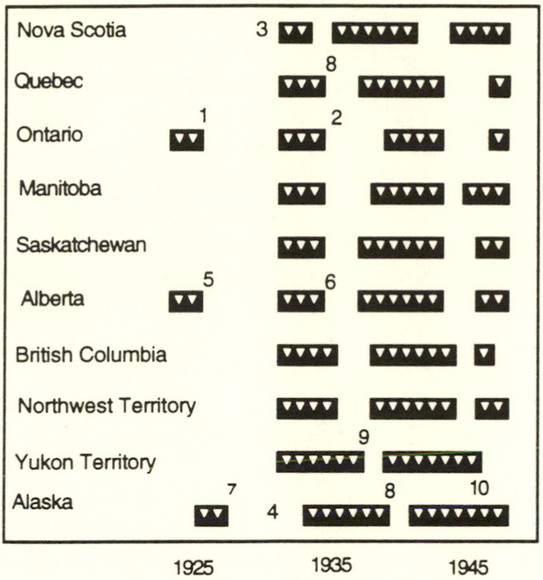

Figure 10.2. Asynchrony between east and west. Black strips show reports of increase or peak abundance among snowshoe hares. Each inverted triangle represents the majority opinion for that year. Numbers refer to events described in the notes. Reports from New Brunswick were too fragmentary to be plotted. (From H. Chitty (1950a) and earlier papers).

At the same time, 1938–1939, 89% of observers reported recovery in Nova Scotia, fur farmers bought large numbers of hares to feed to mink and silver fox, and in the next year hundreds were shipped alive from New Brunswick to the U.S.[22] Then in 1941 came the crash followed by scarcity. Meanwhile, in central Canada and on the prairies hares were recovering their numbers and did not crash until 1943, those in northwestern Canada crashed 1–3 years later still, and those in western Alaska were still abundant in 1947.

Although Figure 10.2 suggests that the cycle progressed tidily from east to west, MacLulich concluded otherwise for the period 1930 to 1934. He writes (p.26) that "abundance was reached and passed first in the Maritimes, southeastern Ontario, southern British Columbia, and the Mackenzie delta, and occurred last in the interior of the continent . . . and in Alaska and Yukon."

Our own enquiry suggested that the length of the cycle had been more than 10 years in Alaska and less than 10 years in Nova Scotia, where, especially on Cape Breton Island, it had had definite peaks in 1931–1932 and 1939–1941, and where hares were extremely plentiful in 1947–1948. In Alaska the peaks seem to have come at longer intervals: in 1925–1927 and in 1937–1938 (in places) and some time after 1947–1948, when hares were still abundant. As a result, cycles at the extremes of the continent, where they had been out of phase since the 1920s or before, were coming into phase, at least temporarily. Local differences are too marked, however, and data are too scrappy for regional differences in length to be established from present data; for even within regions, cycles vary from 7 to 12 years,[23] which makes questionnaires unreliable for determining differences in length.

A new and improved long-term enquiry is worth considering. It would be worth knowing whether cycles are still out of phase across the continent and whether they really are shorter in the Maritimes than in Alaska. If, as seems likely, the effects of weather are strong enough to produce synchrony, they may be strong enough to affect the length of a cycle.[24] A new enquiry, if undertaken, should be confined to perhaps a dozen biologists in different regions. One reliable report is worth more than several doubtful ones, which must nevertheless be included.

It is unfortunate that none of the eight authors of a 1993 paper[25] checked the reference to the author they quote for their belief that "in general the cycle was synchronized over the whole of Canada and Alaska within 1–2 yr." In fact, the author[26] showed that cycles varied *within Canada alone* from over 2 years before the 'national mode' to 2 years after; and if reports from Alaska are included, populations have been as much out of phase as is possible for a 10-year cycle. These authors' analysis of the sunspot and snowshoe hare cycles confirms Whitehead's dictum that "If only you ignored everything that refused to come into line, your powers of explanation were unlimited.[27]

10.3 The Leslie Model

Like other exploratory processes, [scientific method] can be re-
solved into a dialogue between fact and fancy, the actual and possible; between what could be true and what is in fact the case.
                                                        Medawar[28]

Until we can explain fluctuations in numbers we should perhaps leave the problem of synchrony on one side; but an explanation that kills two birds with one stone has much to recommend it, which is another advantage of explaining cycles in terms of qualitative changes. Such changes imply that populations vary cyclically in their response to weather. It is also of historical interest (at least to me) that the need to explain synchrony first made me realize that the density-independent action of weather is a figment of the imagination. Contrary to the misunderstanding of Hansson and Henttonen,[29] qualitative changes associated with self-regulation render populations highly responsive to extrinsic variables, especially those associated with geographical differences;[30] they are therefore of unpredictable length.

The notion that weather kills the same proportion of a population whether it is abundant or scarce depends on the argument that because weather *itself,* unlike (say) predation, is unaffected by population density, its *action* is also unaffected. On the contrary, its action, like that of drugs, poisons, disease, and other extrinsic factors, varies with the susceptibility of the animal and can be known only from sense data, not from reason alone. The unstated assumption in much population theory is that individuals are equally susceptible at all population densities, an assumption that is certainly untenable for populations with different age structures.

If two populations are out of phase, they will presumably remain that way unless some external agent enables one to catch up with the other. If weather kills the same proportion in each population, it cannot do the trick. But as weather is the most plausible synchronizing agent, we must imagine it acting differently on two such populations. One way for weather to act would be for it to kill all animals unable to find shelter in a limited number of protective niches. More would then die in the larger of two populations living in similar habitats. For voles, the evidence against this view is that, despite the variety of habitats sampled by our long trap lines and trace lines, populations disappeared at the same time regardless of habitat type. Also, for a more general solution we need one that applies to relatively uniform habitats and to species other than voles.

To see if George Leslie could give mathematical respectability to a mere verbal model, I kept twisting his arm until he got to work on the idea that populations that differ in quality would be synchronized by highly variable weather. He first imagined a population consisting of four age groups, each with its own rates of fertility and mortality.[31] When subjected to various mathematical operations this population fluctuated, though with decreasing amplitude. Even George's genius was unequal to constructing anything better than a damped oscillation for this kind of population.

Two such deterministic populations, if started out of phase, would remain that way in a constant environment; so the next step was to imagine them in two separate environments with the same weather. George then chose five kinds of 'weather,' varying from very good to very bad, and applied them in random order to two out-of-phase populations. He started each population at about the same density (Figure 10.3), one taken from before and one from after a peak. As a result, they differed in quality and therefore in the way they responded to "weather."

Population A continued to increase at first and population B to remain low; but shortly thereafter population A declined abruptly from high numbers and became as vulnerable to the weather as population B. They then fluctuated in step, with the oscillations becoming damped. It seemed possible, however, that if the weather had been good instead of bad, as it was at the time of the third peak, it might have prevented the oscillations from becoming damped. With a modeler's Zeus-like control over the weather, George therefore let it be good and showed that the resulting peak (C) would have been much the same as in the previous cycle. Under sufficiently variable weather, therefore, the oscillations might have continued for some time before becoming damped.

The tendency for such oscillations to become damped is not what happens in nature, where conditions vary from season to season and year to year. Such variations may be necessary for the persistence of natural cycles, and the attempt to keep a credible model from petering out may be doomed to failure in a constant universe.

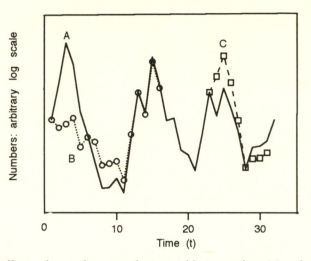

Figure 10.3. Effects of a random 'weather' variable in synchronizing the numbers of two populations, A and B, that were out of phase at $t = 1$. After $t = 10$ their numbers were indistinguishable. The effects of 'good' instead of 'bad weather' at $t = 22$ are shown by curve C. From Leslie (1959).

George's model and that of Moran[32] are only two of an infinite number of possible models and like all models are of limited scientific value (as George took pains to point out) until the dialogue is further advanced "between what could be true and what is in fact the case."[33] It nevertheless boosts morale to have mathematicians find a grain of truth in one's idea; but there's no reason to be downcast if they prove to their own satisfaction that your ideas won't work.[34] Most know as little about biology as biologists know about mathematics. Being able to understand the language of the Ultimate Mathematician[35] guarantees no access to His secrets. Empiricists should take comfort from the fallibility of Lord Kelvin, no slouch as a mathematician, who received no inside information about radioactivity and its warming effects. His estimate of the rate of cooling of the earth, and hence its age, was therefore wrong by several orders of magnitude.[36] His logic ran a poor second to the observations of the imprecise geologists and biologists. Once more the lesson for the field worker is to seek simplicity, this time of the mathematician, but to mistrust it. (Whitehead, again!).

Review, 1923–1961

. . . It is harder to discover the elements than to create the
science.

Whitehead[1]

It should be the chief aim of a university professor to exhibit himself in his own true character—that is, as an ignorant man thinking,
actively utilizing his small share of knowledge.

Whitehead[2]

11.1 Discovery

If scientific method included a recipe for making discoveries (as opposed to
piling up data), this book would have been shorter. Anything resembling
method, however, is restricted to the time after one has made a discovery, or
thinks one has done so. Only then can something methodical be applied to
see if there's anything to it. This is a far cry from the popular idea of scientific
method. According to a distinguished classicist,[3] for example, "the scientist is
no doubt rigidly objective: he collects the facts relevant to a problem and
makes no conclusions that the facts do not justify." If this were true, scientific
research would be impossible. For there are no *a priori* means of knowing
which so-called facts are relevant to a problem, nor can there ever be enough
of them to justify a universal conclusion, a problem dealt with in the next
section, nor can science proceed without risky leaps of the imagination. The
present section deals with the long and difficult prerequisite to anything resembling the cold logic often supposed to characterize scientific procedure.
The imaginative element,[4] is not a product of method any more than it is in
the arts.

The present story starts in 1923 with Charles Elton's tales about lemmings
in Norway and comes full circle in 1959 with Charley Krebs's study of lemmings on the west coast of Hudson Bay. In between are 30 years' work on
the short-tailed field vole, whose study, it was assumed, would lead to the
discovery of the elements responsible for population cycles in voles, lemmings, and perhaps other species. This expectation has not been fulfilled, nor
does the view in 1996 look any brighter. "If men could learn from history,"
wrote Coleridge,[5] "what lessons it might teach us! But . . . the light which
experience gives is a lantern on the stern, which shines only on the waves

behind us." That may be all the present story will accomplish, since the two main lessons spelled out in 1942—to mistrust mere correlations and look in new directions—have still to sink in.

Explanations in this prewar work were first assumed to be consistent with a process in which populations, whether of mice or moose, suffered increasingly heavy losses as density increased. The animals were supposed to die from starvation, predation, parasitism, or disease, the latter being the favored explanation for small mammals. Unlike weather, these mortality factors are capable of responding in a density-dependent manner, which is assumed to exert sufficient feedback to prevent unlimited increase in numbers. In other words the causes of death are both an effect of increasing numbers and a cause of bringing them to a halt. Many population ecologists think that establishing density-dependent relationships is sufficient to explain the regulation of numbers. That correlations are worthless without controls is one of the features that makes the critics of population eology doubt whether it can be regarded as a science. We should not be blind ". . . to the limited explanatory value of studies that are not designed to distinguish between causes and effects." [6]

Evidence in favor of this reigning explanation is that vegetation is often damaged, predators and parasites become obvious, and disease is sometimes prevalent. But although such changes in severity are almost bound to be an effect of rising numbers, they may be no more than that. Some of us—John Christian, David Davis, Paul Errington, myself, and perhaps others—therefore rejected these obvious answers before or soon after World War II and concluded that no solution to the problem was credible unless the animals' behavior and physiology were also taken into account.

Faced with evidence for a rival point of view, "a scientific community can (a) ignore or dismiss it, (b) accept it but deny its relevance to existing theories, (c) reinterpret it to leave existing theories intact, or (d) use it to modify or reject existing theories." [7] Any of these reactions are justifiable at different stages of one's own research or knowledge of other people's. David Lack and I disagreed because he chose options (a) and (c), whereas I chose (d).

In work on the vole, the problem that first received attention was the decline in numbers. Until this was explained, two other problems were unlikely to be solved, namely, the tendency for these losses to recur every 4 years and to be synchronous among independent populations of the same and other species. The most reasonable explanation for the decline was the epidemic hypothesis. But in questions of biology, as Huxley observed, "If anyone tells me "it stands to reason" that such and such things must happen, I generally find reason to doubt the safety of his standing." [8] Unfortunately, notions that stand to reason are hard to get rid of, and it took the 10 years 1929–1939 to decide that *changes* in the obvious sources of mortality, alone or in conjunction, were insufficient to explain the decline. Neither was a change in any of them necessary, as some declines took place among disease-free voles that apparently had plenty of food and few enemies. Nor could bad weather be blamed for the declines. So what were the animals dying from? There seemed at first to be nothing left.

An alternative approach was suggested by two pieces of prewar work. Two of my colleagues suggested that a declining population consisted of voles that were older than usual and correspondingly more likely to die of anything whatever. My own work, on the contrary, suggested that the individuals that disappeared were young but congenitally abnormal, owing to the stress of competition for space among their mothers. Although both explanations were wrong,[9] both implied that the original question should be changed from "what kills the animals?" to "what makes them more likely to succumb to chronic disease, bad weather, and other local hazards?" This way of rephrasing the question has two advantages: it gets around the difficulty that mortality factors differ from time to time and place to place, which means that one must seek elsewhere for antecedents that are common to cyclic declines, and it provides a means for numbers to be synchronized through random changes in weather. Thus, weather, though itself unaffected by population density, might nevertheless affect some kinds of populations more severely than others. If so, its *action* would be density dependent, and weather could be brought back into population theory, from which it had been banished.

Given that feed-back of some sort is necessary for the regulation of numbers and that the extrinsic environment remains tolerable, we must postulate some change in the animals themselves. Figure 11.1 sums up the difference between the conventional wisdom (A) and the new heterodoxy (B).

According to A, density-dependent changes in extrinsic variables—food supply, predation, parasites, and disease—are both necessary and sufficient to prevent further increase or start a decline in numbers. According to B, the only necessary change is in the quality of the animals, which, according to A remain constant (though this is seldom stated explicitly), and the course of

Figure 11.1. A, Hypothesis that populations are regulated by density-dependent processes. B, Hypothesis that populations are regulated by behavioral changes and natural selection.

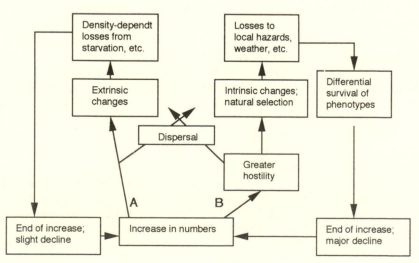

the cycle is determined by hostility early in the increase phase. Changes in susceptibility in conjunction with purely local hazards are then sufficient to regulate population density—but not when these hazards are reduced in the field (for example, during unusually good weather) or eliminated in captivity. Thus, according to this scheme, local mortality factors are necessary, but density-dependent changes in their severity are not, though they may be associated.

This scheme implies that causes of death, being unpredictable, do not warrant the care so often lavished on correlating them with changes in population density.[10] So-called predictive equations, based on natural history of the past, will not necessarily predict the future. Knowing the enemies of the winter moth in Wytham Woods, Oxford, was of negligible value in predicting the course of events in Nova Scotia and British Columbia.[11]

As I now saw the problem, we needed to know, for cyclic and noncyclic populations alike, what prevents unlimited increase in abundance among populations with plenty of food and few enemies. I had come to view both kinds of populations as variations on a common theme of self-regulation of numbers. The great advantage of studying cyclic populations is that one can make comparative observations between increasing populations (the control) and stationary or declining populations (the phenomenon to be explained). Opportunities for studying an unregulated population crop up rarely and unexpectedly among noncyclic populations, and their importance may go unrecognized (as it was after the Great Tits in Wytham Woods recovered from the hard winter of 1947).[12]

The concept that a difference in quality is the essential difference between increasing and regulated populations has remained intact, but its explanation has suffered several sea changes. After dropping the idea that changes in quality are purely maternal, I proposed that all populations are capable of getting rid of surplus animals (or limiting immigration) and will remain abundant or decline only slightly when selection is absent or slight; second, that populations become cyclic when subjected to rapid natural selection between two or more behavioral morphs. (The system has to be simple or it won't work quickly enough.) In voles, I believed the increase phenotypes to be docile and the decrease phenotypes to be aggressive. On balance this seemed the most plausible reason for the good survival, growth, and reproduction of the former and the success of the latter in winning out during peak and decline. The only snag was that the phenotypes born during the increase phase were larger and more likely to beat up the smaller "decrease" phenotypes. I had to explain away this awkward anomaly by assuming that the latter were the microtine equivalent of the lean and hungry men that Julius Caesar found so dangerous. A quarter of a century later I got round to testing behavior and, with the help of an undergraduate, Sharon Barton, discovered that I had things backwards[13] and that the large animals were in fact the more aggressive. (My colleague Robin Liley had told me this all along, but enmeshed in the toils of my own logic, I chose not to believe him. Zinsser[14] has likened this state of mind to that of hens sitting on boiled eggs.) From this and other

evidence[15] I then decided that the apparent docility of the increase phenotypes was due to their breeding amicably within well-defended and spatially
distinct family groups (which is compatible with the idea that increasing populations of voles[16] and grouse[17] tend to consist of closely related individuals
and that "some animal populations comprise distinct subpopulations or
demes").[18] If this social organization breaks down at peak densities, so ran
the next version,[19] it would be an advantage for young animals to avoid
conflict by spacing themselves out and failing to mature early. "Becoming
sexually mature in an established population may be extremely hazardous."[20]

To return to the historical sequence: the postwar work consisted, first, of
replicating the prewar work, second, of experimental work in search of a
mechanism to explain a supposed change in susceptibility. High density
alone[21] was not enough to produce a decline; without some lag effect it could
not explain the persistent fall in numbers and refractory low phase. But critics
were unhappy with the idea that competition for space might set a limit to
peak numbers and were justifiably skeptical of the idea that its further effects
could explain the long decline or low numbers. Though wrong[22] in suggesting that the viability of later generations was reduced by maternal stress,
this hypothesis eventually confirmed statements by authorities—from Bacon,
Bernard, Huxley, and de Morgan to Popper—that even wrong ideas properly
disciplined will get you somewhere in the end. Meanwhile, none of the evidence from adrenal weights, liver glycogen, blood, or stress in the lab supported the original view that voles were nonviable.

The explanation for the decline now shifted back to the notion that in
spite of evidence to the contrary the environment was indeed unfavorable
during the decline. The element missing from the standard environmental
picture seemed to be the kinds of neighbor surrounding each individual. By
the late 1950s intraspecific strife had gained recognition as a factor associated
with high numbers, and a few people were willing to believe it left a physiological legacy. Evidence now suggested that the legacy was primarily behavioral and that the voles that disappeared were the victims of an environment
that they themselves had made unfavorable. This idea was supported by the
discovery that the decline was even more abrupt than previously suspected.
Also, as sex and age groups disappeared at different times, the observations
were more simply explained by changes in behavior than in viability or the
severity of mortality factors. Behavioral changes, whether in threat or avoidance, that drove animals out of their normal runways would explain an increase in susceptibility to predation, which might sometimes be the com-
monest cause of death (and thus lend spurious confirmation to the belief that
predation is sufficient to regulate numbers).

By concentrating on peak and decline, we largely neglected the other two
phases. Studying the low phase was hampered by working at a distance,
which made it expensive and time consuming to get adequate samples from
almost nonexistent populations. We knew only that growth, survival, and
sometimes reproduction were less satisfactory than during other phases. Nor

do we yet understand the difference between low-density animals at the end of a decline and those at the start of the increase phase. The latter, in contrast to the former, may begin the season early and end it late, produce offspring that survive and grow well, and themselves grow well in the lab. Such phenotypes may be necessary for cyclic behavior. Reproduction in the continental vole is exceptionally good during the increase phase,[23] and we ourselves had lab evidence that litter size might be increased by mild interactions with neighbors. As our lab voles lacked such stimuli in earlier work, their intrinsic rate of natural increase may have been underestimated, especially as they may also have received a slightly toxic diet.

By 1957, a profitable mistake had shocked us into rejecting maternal stress as a plausible mechanism for transmitting the effects of crowding to subsequent generations. An alternative was that the selective advantage of a high reproductive rate and certain kinds of behavior might vary with the phase of the cycle: one kind of young might be favored when there's no competition for space and a different kind might be favored when there is. This view was supported by the discovery that declining populations consisted of a mixture of animals. When removed to the lab, some became as heavy as peak-phase adults, others remained light in weight however long they lived. The latter were the main survivors in the wild. Instead of being nonviable, they must have some selective advantage[24] (shades of David Lack!); so for many years I held the counterintuitive idea, mentioned above, that these small morphs were more aggressive than the large ones present in the peak.

Throughout these years we did what we could to see if explanations for the vole cycle also explained other cycles. The 4- to 5-year decline in the snowshoe hare is the most powerful argument for inventing hypotheses to explain the lag effects of hostility. But no explanation can be considered satisfactory that fails to explain the tendency for cycles to be regular and synchronous within and between species. Variable weather is the most likely synchronizing agent, either directly or indirectly, though its action has yet to be studied. Explanations are also suspect that ignore the changes from high body weight at the peak to low body weight after the decline. This feature is still unexplained. It's the best evidence that changes in quality are associated with the mammalian cycle, but is only one symptom, I assume, of qualitative changes expressed in various ways in other species. An implication of this hypothesis is that in unfavorable habitats population density will be kept below the level at which the animals compete for space. It should therefore be possible to prevent cycles by cropping a population hard enough to keep it in the increase phase. Mike Smyth, who first tried to imitate the effects of heavy predation on bank voles, found that until late winter so many surplus animals were wandering around that he was unable to keep them from pouring into his unfenced experimental plot. His results were consistent with the idea that all species are capable of regulating their own numbers and producing a surplus of young doomed to die from predators and other agents. But owing to immigration and the absence of a cyclic decline the experiment failed to test the lag effect. Mike nevertheless pioneered the way for later manipulations (and incidentally showed how intense predation would have to be to have much effect on the density of a prey population living in a favorable habitat).

Implicit in ideas about vole populations is their relevance to the species whose alleged suicidal behavior started the whole enterprise. Charley Krebs found that the Canadian lemming cycle ended without the mythical mass migration, though something like it had happened during the peak. He also found that symptoms of the decline were the same as in voles, in particular, that body weights were high in peak populations and low during the decline. These and many other findings encouraged him to spend his career in trying to explain them. He's still trying.[25]

My own experimental work now stopped for over a decade. At my last scientific meeting before leaving England I explained that "the hypothesis being examined was that a decline in numbers is a necessary consequence of the selective action of hostility against the genotypes present during the phase of rapid increase."[26] This statement is logically equivalent to saying that selection is *sufficient* to produce a cycle, which I did not realize and did not believe. Fortunately, no one seems to have spotted this inconsistency with my claim that selection is merely necessary. Only if the *ceteris paribus* clause is satisfied can hostility be said to be a sufficient condition for cycles. But as selection may have too little diversity to work on and as other variables are also uncontrolled, it's easier to protect the proposition that selection is sufficient for the occurrence of a cycle than that it's necessary. Hypotheses are supposed to be tested, not protected.

Putting these ideas into better shape took until 1967. The Achilles heel of the concept is still the unlikelihood that natural selection can act as fast and as strongly as required. If it were not for the difference in time scale, the concept would be a special case of the r/K selection ideas of MacArthur and Wilson,[27] which were also published in 1967. Both schemes assume that individuals in crowded and uncrowded environments will be selected differently. In colonizing an island or recovering from scarcity, some phenotypes will be better than others at filling the empty space. Such individuals are said to be 'r-selected' (r being the symbol for the intrinsic rate of natural increase). But as numbers increase and approach an asymptote K, other phenotypes will be better at passing their genes to later generations. Such phenotypes are said to be 'K-selected.' Individuals in the early increase phase of a cycle can thus be regarded as subject to r-selection, those near and at the peak to K-selection, and those in the decline and low phases as survivors of the latter. Even if false, the idea is worth testing; and if voles had been as easy to work on in Vancouver as they were at Oxford (or if I had not switched jobs) I might have got to first or even to second base. As in other branches of zoology, much depends on you and the right animal being together in the right spot. (Despite what cynics say, going to exotic places such as the Caribbean is often the best way to solve a biological problem.) My preference now would be to use insect populations for testing ideas about qualitative changes; if the ideas can be rejected on evidence from insects, they are unlikely to apply to mammals, at least in their present form. You can't have one without the other. The author of a promising rival scheme agrees that "population quality has been slighted" but believes that "The impact of maternal effects on population quality can, on its own, generate population fluctuations.[28]

My own faith in natural selection as a necessary but not sufficient condition is bolstered by the following argument:

> . . . selection arising from competition does not necessarily make succeeding generations of individuals of that species more fit to meet circumstances other than those of the competition itself. Selection arising from competition will favor nothing but the ability to survive that particular type of competition. Indeed the selective rise of a special type of competitive power may be accompanied by a reduction in other components of fitness.[29]

To sum up: (1) population density is only one of many environmental variables affecting population trends and by itself is almost useless for predicting them or the behavior of the individuals. (2) Differences in extrinsic variables may explain differences in abundance between habitats, but vary too much from place to place to explain differences between regulated and unregulated populations. (3) We need some measure of 'crowding' that takes account of the kind as well as the number of individuals and recognizes that high and low population density are relative terms that are not comparable between habitat types.

We now turn to the problem of testing the conclusions so laboriously arrived at.

11.2 Justification

(1) It is easy to obtain confirmations, or verifications for nearly every theory—if we look for confirmations. (2) Confirmations should count only if they are the result of risky predictions. . . .

Popper[30]

Sherlock Holmes' famed "powers of deduction" were really powers of making shrewd inductions.

Hospers[31]

One does not, as writers of detective stories seem to imagine, deduce hypotheses; quite the reverse, hypotheses are what we deduce things from. It was at one time thought that hypotheses could be arrived at by a rigorous logical process of 'induction'.

Medawar[32]

Though unsolved in the 73 years since 1923 and of limited interest to most of mankind, the cycle problem forced us to consider a number of methodological problems that should concern all population ecologists as well as other scientists. For though 'scientific method' provides no method for making discoveries, it insists on their having logical consequences that in principle are universally testable. It has been said that "a first-rate theory predicts; a second-rate theory forbids and a third-rate theory explains after the event." [33] According to Popper,[34] however, the more a theory forbids the better it is; a first-class theory must therefore comprise both positive and negative predictions; a second-rate theory tells *only* what is not. Predictions must, of course, be sufficiently precise to be falsified: vague predictions don't count; they are easily verified.

These criteria are not easy to apply. Ideas I long ago considered incompatible with the evidence are still going strong and are likely to remain popular until replaced by one that works. By an idea that works, I mean one that meets standards acceptable to the rest of the scientific community rather than one that's tailor-made to protect the idols of the ecological market place.[35] Rightly or wrongly I had come to regard current theories of density dependence as comparable to Aristotelian physics—excellent as a rational explanation for past events and protected from refutation because of the vagueness of their predictions. The message I've tried to put across is that progress depends on one's predictions being refutable so that one can advance in small, manageable steps. I've been right, and I've been wrong, and being right is better—but not much better if no one thinks your ideas worth testing. It's said that "no one believes an hypothesis except its originator but everyone believes an experiment except the experimenter." [36] He alone knows the trouble he had to make things come out right. Nevertheless, time the great healer removes these doubts, and experimenters are no less vulnerable than theorists to becoming mesmerized by untested or unreplicated results.

Let us now suppose that many years hence you think you've solved the cycle problem. Suppose that you've weeded out all known explanations but one, that it follows from plausible premises, is consistent with a wide body of theoretical knowledge, has led to new discoveries, makes precise, repeatable, and novel predictions, and has survived attacks by its critics. Your conclusions will indeed be based on the best evidence available in this imperfect world, but in spite of their credentials, will not have been proved. This is no disgrace, for being scientific, they never can be proved, and you may take comfort from Bertrand Russell, who wrote as follows:

> Suppose . . . that you invent a hypothesis, according to which a certain observable quantity should have a magnitude which you work out to five significant figures; and suppose you then find by observation that the quantity in question has this magnitude. You will feel that such a coincidence between theory and observation can hardly be an accident, and that your theory must contain at least some important element of truth.[37]

Russell nevertheless warns us not "to attach too much importance to such coincidences." Scientific conclusions do not follow from a pure deductive argument, which alone can guarantee proof. Proof is restricted to an abstract world, whereas natural science deals with a real world less amenable to "fantasies of wish-fulfillment." [38] It's easier to play around with propositions than with animals, which accounts for the high proportion of population biologists who restrict their talents to modeling. While unwise to reject bright ideas from the Olympian thrones of their armchair colleagues, gumboot ecologists should recognize that theoretical conclusions are true only for an ideally isolated universe. As Whitehead is their witness, if anything at all is out of relationship, one's ignorance is complete.[39] You cannot make bricks without straw. Ecologists have been slow to provide the straw and theorists overzealous to provide the clay. Some of their structures are correspondingly friable.

If science depended on proof rather than justification it would never advance as it does through the sequence of provisional conclusions by which knowledge becomes established. But because a wrong conclusion can involve one in years of irrelevant work, the best insurance against this fate is to have critical colleagues, and students who are encouraged to bite the hand that feeds them. My own hand is covered in scars. Population ecologists also need insurance through something equivalent to Koch's postulates,[40] which were designed to prevent bacteriologists from jumping to conclusions based on correlations, which is what so many ecological explanations are based on.

It can be argued at this distance in time that if the epidemic hypothesis had been looked at more critically—if Koch's postulates had been applied—it might have been abandoned sooner than it was. This would have saved thousands of hours of work that threw no light on reasons for the decline. Nevertheless, if the epidemic hypothesis had been given up any sooner, Dr. Wells would not have joined the Bureau and so would not have discovered vole tuberculosis. Nor can we guess what else we might have missed, for refuting this hypothesis led us to the maternal stress hypothesis, which led us to study physiology and behavior, which led us into population genetics, which led us to try to integrate population ecology and population genetics. As with the house that Jack built, one contingency led to another; indeed, in the pursuit of knowledge one rarely follows the original track. As well, if shock disease had gripped the imagination less strongly between 1937 and 1957, it might also have freed us to follow more fruitful lines. We shall never know. The difficulty was to get evidence against it that was more persuasive than evidence for it. Indeed, as is commonly the case,[41] its nemesis was a better alternative rather than evidence against it. The strength of pure research lies precisely in this freedom to change course as it opens up more fruitful lines of enquiry. Applied research is more difficult and sometimes less fruitful.

Although free to change direction, most of us pick problems we wish to solve and, in graduate school especially, cannot risk getting stuck down blind alleys. We must close our ears to the siren song of merely favorable evidence, which can lure us into the fallacy of affirming the consequent, that is, inferring the truth of one's hypothesis from the accuracy of its predictions. The operative word is *truth*. Scientists do in fact affirm the consequent all the time but unlike logicians are free to accept or reject its implications. Logic must be tempered with wisdom, or what passes for it at the time.

If the Scylla in science is being lured to one's doom by deceptively favorable evidence, the Charybdis is being too easily discouraged,[42] "for we are all likely to make mistakes except those of us who do nothing."[43] The course to steer is somewhere in between; one should look for contrary evidence, make a note of it in accordance with Darwin's Golden Rule, but not abandon one's hypothesis until something better comes along. In the meantime, finagling is not uncommon even among the best scientists and is less reprehensible than it seems.[44] Compared with the convolutions of some authors, my own finagling looks like the work of an amateur.

The traditional safeguard against mistaking correlation for causation is to make comparative observations. Comparative methods are most familiar in the form of comparisons between experimental and control observations, for which population ecologists frequently substitute long-term descriptive studies in one and the same place. By themselves, such studies provide too few opportunities for refutation, and conclusions drawn from them must be viewed with suspicion. What appears to be true for a particular population can be accepted if and only if it applies more generally. The explanation is otherwise false or parochial.

The prewar work of the Bureau comprised enough different areas that explanations that were true to the evidence in one place were patently false elsewhere. Some inferences did indeed depend on judgment rather than quantitative data; also on faith that declines are a recognizable phenomenon. Even at this preliminary stage, however, the principle was to look for what Bernard calls 'counterproof,' and to reject ideas on finding an event from which the supposed cause was absent. (Symbolically, CE must be confirmed by C'E', namely, by showing that when C is removed or absent, E is also absent; hence, that CE is no mere coincidence.)

To a zoologist, comparative methods are also those used in drawing inferences from similarities and differences among species. In doing so, one runs the risk of making a false analogy and viewing a world distorted by delusions of grandeur, as one's critics never fail to point out. This is a professional hazard. But taking risks is almost always better than protecting one's hypothesis indefinitely—a fault that must be distinguished from sticking to an idea that's hard to test. If it's right, there's nothing to worry about; if it's wrong, it's better to find out for oneself; at worst, one learns from criticisms provoked by a stimulating error. Nothing venture, nothing win.

Ideas worked out on one species are unlikely to apply to others in every detail. It's therefore wise to distinguish between essential and non-essential or accidental characteristics. This is the advantage of hypotheses based on the theory of natural selection: by cutting across taxonomic boundaries they direct one's attention from specifics to essentials and thus increase one's chances of seeing what's wrong with them. The only danger is that theory, if wrongly applied, may discourage experimental testing, for "the intolerant use of abstractions is the major vice of the intellect,"[45] as we have seen with the theory of density dependence.

Because we can all cook up evidence in favor of our pet hypothesis (I have frequently done so), scientific philosophy adopts the heuristic device of assuming its opposite. As this means that you and your critics are united in a presumably objective test of the same null hypothesis, scientific controversies should not drag on as long as they do. Unfortunately, the so-called crucial experiments are not that crucial, as they are never 'theory-neutral' and so are usually interpreted according to the preconceived notions and vested interests of the opponents (for we are all less objective than we like to think). Also, "It is very difficult to defeat a research program supported by talented, imaginative scientists."[46]

In the present controversy, the most appropriate null hypothesis is that there are no heritable differences in behavior between animals from increasing and other population phases. If accepted, this null hypothesis destroys the idea that animals can regulate their numbers; if rejected, it confirms that such behavior is possible; but instead of being a necessary condition for cycles, genetic change may be merely an effect of demographic change.[47] I assume it's both: that selection is due to demographic change and that further demographic change is due to selection. Testing the null hypothesis is only the first step, however; the experimental hypothesis must give positive results: it must show how to produce cycles artificially and how to suppress or otherwise manipulate phases in natural populations. Experience shows how difficult this is.

Population ecology is likely to rely for years to come on observation rather than experiments of traditional form. According to Bertrand Russell, "in an experiment the circumstances are artificially simplified, so that some one law in isolation may become observable."[48] Fortunately for ecology, this is no longer the definition of an experiment, there being far too many uncontrollable variables in the field. In the modern view, an experiment is a test of a hypothesis,[49] evidence against it being logically equivalent whether based on natural or manipulated events. In spite of this broader definition, however, the difficulty of doing experiments in population ecology—even by the few authors dedicated to this approach—is another reason for the coexistence of so many competing research programs. There's no incentive to think up new solutions as long as people cling to the old ones through accentuating the positive instead of eliminating the negative.

Even contrary evidence is inconclusive, however. Unlike the abstract world of logic, where a single instance can falsify a general proposition, science never proceeds in isolation from the real world. Every hypothesis is tested in conjunction with assumptions and auxiliary conditions known as the *ceteris paribus* clause. So it's always possible to blame contrary results on faulty experimentation or uncontrolled variables or to ignore, deride, dismiss, or somehow reconcile them with conventional wisdom. As well, many so-called facts turn out to be anything but, and many of the brightest ideas burn out like Roman candles. For in population ecology plenty of things can go genuinely wrong and justify an *ad hoc* hypothesis. But ideas soon lose credibility if they have to be protected too often.

Failure to understand that scientific conclusions cannot be proved has a serious practical consequence. People whose vested interests are threatened can claim, correctly, that there's no 'scientific proof' that their products are harmful. Thanks to this ploy, remedial action has been delayed on tobacco, pesticides, acid rain, chlorofluorocarbons (CFCs), and other destructive products. For example, in 1975 the DuPont Company defended CFCs on the grounds that "to date there is no conclusive evidence to prove" that they "will cause a health hazard by attacking the earth's ozone layer."[50] Though correct in claiming that "much more experimental evidence is needed to evaluate the ozone depletion theory," the company thought it irresponsible to act on the existing evidence. The fallacy that lack of 'proof' permits one to discard judgment is one of the legacies of the popular misunderstanding of science.

The opposite misconception about science is to assume that conclusions in fields other than science can be proved through applying scientific method. The historian Arnold Toynbee, for example, justifies his poetic visions on grounds that are in fact "a travesty of the scientific method," for "nothing is more likely to be misleading than the comparison of the historian's method with that of the scientist."[51] Nor are scientists best qualified to solve the political, cultural, ethical, religious, and esthetic problems of mankind.

The cycle problem will be solved one day, perhaps when conclusions so far reached have been rejected in whole or in part and the decks are clear for new and imaginative solutions. A single paragraph in the text books may condense the work of a century or more. But of more lasting significance is the lesson to be learned from this long lapse of time, namely, that "The raw materials out of which science is made are not only the observations, experiments and calculations of scientists, but also their urges, dreams and follies."[52]

In one sense, the present account is a story of failure. Medawar writes:

> No scientist is admired for failing in the attempt to solve problems that lie beyond his competence. The most he can hope for is the kindly contempt earned by the Utopian politician. If politics is the art of the possible, research is surely the art of the soluble. Both are immensely practical-minded affairs.[53]

If taken at face value, this statement would hold Harvey in contempt for failing to solve the problem of respiration and warm-bloodedness, everyone but Watson and Crick[54] who tried but failed to crack the genetic code, and all who pursue a line of research that stretches beyond their own careers. Such a casualty list would already include a distinguished company of people who have worked on cycles—a company that's likely to be swelled by new members until well into the next millennium. When read in context, however, Medawar's warning is clearly directed against attempts to apply scientific method improperly or to problems for which it is unsuited. There's no reason to assign the problem of cycles to such a category of lost causes, in spite of its origin among the dreaming spires of Oxford. But we may well be concerned at its tortoise-like rate of progress, which prompted my colleague Peter Larkin[55] to write the following obituary:

Dr. D.H. Chitty was this morning found drowned in one of the fish tanks belonging to the Department of Zoology, U.B.C. There was no evidence of foul play and it is feared that Dr. Chitty took his own life. The only clue to what one person called a "tragedy" is a clipping from the *Province*, entitled "Army of Mice," which was found clutched in the dead man's hand. Across the top of this clipping was an enquiry from the Head of Dr. Chitty's department saying "How come?" and beneath it was written what are believed to have been the dead man's last words: "I don't know."

The sudden realization that people expected him to know the answer to the problem on which he had spent his entire professional career is believed to have unbalanced his mind.

Dr. Hoar expressed his acute distress at the incident. He said that Dr. Chitty should have known better than to risk disturbing the fish.

It's lucky that finding an answer is not the only measure of achievement in pure science. More professors would otherwise copy my alleged lemming-like behavior.

Let us now suppose that instead of hiring me in 1935 Charles Elton had been lucky enough to hire a reborn Sherlock Holmes who had studied population ecology at Oxford (or was it Cambridge?),[56,] and let us ask if Holmes would have solved the cycle problem. My guess is that with his powerful imagination and genius for distinguishing relevant from irrelevant clues, he would have been 50 years ahead of the rest of us. Unlike his brother Mycroft, the prototype of today's modeler, his genius was not confined to armchair speculation but also flourished in "the radically untidy, ill-adjusted character of the fields of actual experience from which science starts."[57] In his D. Phil. thesis Holmes would probably have rejected his early, extravagant claims for the power of logic[58] and would have justified by prediction his conclusion to the Adventure of the Suicidal Lemmings. He would have saved us a lot of trouble.

Epilogue, 1961–1995

Unanimity of opinion may be fitting for a church . . . [but] Variety
of opinion is necessary for objective knowledge.

Feyerabend[1]

One of the strengths of science is that it does not require that
scientists be unbiased, only that different scientists have different
biases.

Hull[2]

12.1 Win Some, Lose Some

*John Clarke was one of my first PhD students, and Rudy Boonstra was
my last. Rudy came to UBC in September 1972 and finished his thesis
in 1976. In an already distinguished career, Rudy and his coauthors
have inflicted one of the scars mentioned in Section 11.2, and by turning
back the wheel full circle to something resembling my 1952 idea (that
declines are due to maternal stress), have risked turning into pillars of
salt. It remains to be seen whether this is the price they will pay, or
whether I shall pay the price that supervisors pay for encouraging students to think for themselves. In the latter case I shall take comfort
from Zinsser, who tells us that: "Our task, as we grow older in a rapidly advancing science, is to retain the capacity of joy in discoveries
which correct older ideas, and to learn from our pupils as we teach
them. That is the only sound prophylaxis against the dodo-disease of
middle age."[3]*

*Difficult though it may be, one should try to retain this capacity,
especially in a slowly advancing science such as population ecology,
where occasions for such joy are rare and must be made the most of.*

Critics of my war cry "no selection, no cycle" rightly doubt whether selection
can act rapidly enough to account for the violent reversals of fortune experienced by cyclic populations. Until 1986, supporting evidence, as opposed to
long-standing suggestions,[4] was limited to showing that gene frequencies of
doubtful relevance can change rapidly during a decline;[5] but in that year
Norah Spears submitted a thesis in which one chapter dealt with the way
voles respond to selection for sexual development in long and short

daylengths. This work, with John Clarke as coauthor, was published in 1988,[6] 2 years after Desjardins et al. had described similar work on deer mice.[7] Both papers showed how rapidly selection can proceed. In the work on voles, selection applied to the F1 generation derived from field animals produced two lines, both recognizable in the F2 generation. One line, the Emergent line, was sexually more developed than the other, the Inhibited line (Figure 12.1). In the work on deer mice, one line became largely infertile after 2 generations of selection under short daylength; the other line was largely fertile (Figure 12.2).

In earlier work on a long-established lab colony, John had failed to select an emergent line.[8] This stock was known to differ from the wild stock: through having been maintained under long daylengths it had presumably lost the phenotypes able to breed under short daylengths.

If animals are selected for breeding in winter in the wild, they can breed only with one another, which would hasten the speed of natural selection and may be one of the conditions necessary for cyclic behavior. I'm indebted to 2 former students, Judith Anderson and Chris Foote, for putting me wise to the probable need for assortative mating.[9]

Winter breeding is typical of increasing populations of voles and lemmings, and as differences in growth rate and behavior are presumably linked to differences in reproduction, we can dimly see how behavior may change throughout a cycle in numbers. Mental gymnastics are nevertheless required to fit this explanation to populations near Berkeley, California, where voles breed every year in winter, and the peculiarity of the increase phenotypes is that they breed in summer. This and other difficulties should not deter one, however, from exploring this new and unforeseen discovery and seeing what comes of it. If past experience is any guide, it will not be the abracadabra

Figure 12.1. Response, under long and short daylengths, of F2 lines of male voles selected for ability to develop sexually under short daylength (E) and for inability to do so (I). Using a 6-hr daylength, Spears and Clarke (1987) selected the two lines from F1 offspring of field animals.

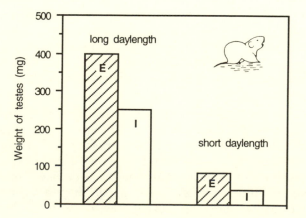

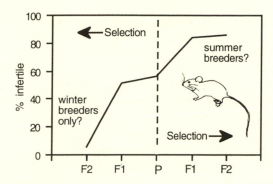

Figure 12.2. Effect of selecting for deer mice that do and do not respond to breeding under short daylength (8L: 16D). P = parental generation. From Desjardins et al. (1986).

some of us think it may be,[10] but it will at least add another piece to the jigsaw puzzle, and even if irrelevant to the cycle problem, will add significantly to general knowledge. Having a strong fall-back position is something one should always consider before starting on a possibly wild goose chase. "After all, why complain if in looking for El Dorado you stumble upon Atlantis instead?"—an appropriate quotation[11] to include in a book about lemmings.

Selection for reproductive prowess is one way of getting round the difficulty that aggression itself may be too complex to be under rapid selection pressure.[12] Moreover, aggression may not be the relevant variable. In its pristine form my understanding of cycles is limited to the claim that "an unknown kind of interaction produces an unknown change . . ." We should not be discouraged by early failures to guess the relevant mechanisms. My own guesses (a.k.a. working hypotheses) have unfortunately been taken more seriously than the methods recommended to test them—in the present case through testing the null hypothesis that there are no heritable differences in behavior between increasing and other phases. A *sine qua non* for such tests is being able, first, to identify the morphs that typify each phase and then to compare and contrast their behavior. An indirect approach to this problem has since been taken by Rudy Boonstra and his colleagues.

Boonstra and Boag[13] launched a sophisticated experimental attempt to test the idea that selection plays a significant part in the cyclic decline of small mammal populations. They used the argument that if natural selection is operating on the genetic composition of a population, "a suite of life-history traits [body weight, growth rate, and age and weight at sexual maturity] is simultaneously undergoing selection and that these traits are strongly heritable." From taking parents from the field in two successive years and breeding from their progeny, they discovered that variation in almost all traits was nongenetic in origin and that "maternal and other environmental effects were of overriding importance." Hence, they doubt whether "individual differences in demographic characteristics are primarily genetic" or that selection would be able to act fast enough to reverse the frequencies of the morphs found in high- and low-density populations.

I myself agree that the traits they studied are strongly influenced by environmental factors; but I disagree that one can infer from studying these symptoms that other processes are unaffected by natural selection. Nor am I convinced by arguments involving so many links in the chain of reasoning. I also believe that some of the discrepancies they found between growth rates in field and laboratory may be explained by the difficulty of distinguishing between the morphs found at the beginning and end of the low phase, namely, those that do and those that do not grow well in captivity.

Mihok and Boonstra[14] compared the breeding performance of voles at the beginning and end of a 1-year low phase in 1985–1986. The 1985 females bred poorly in the lab, but their daughters and granddaughters did well in grassy outdoor pens and poorly in the lab—at the same time that the 1986 females did well. Some of the F1 males failed to grow, others grew normally, varying as adults from 21 to 58 g (Mihok *in lit.*). The authors concluded (1) that as all voles came from low-density populations, they should have reacted in a similar way to space constraints in the lab; (2) that as the males were reproductive in both years, the poor performance of the F1 and F2 daughters was a maternal effect; hence (3) "genetic explanations for the decline (Chitty 1987) are unlikely."

The first conclusion assumes that one can predict behavior from population density, which is contrary to my own experience, and the inference from (2) to (3) strikes me as unconvincing. Furthermore, the results of the breeding experiments, shown in figure 12.3, are just as consistent with selection being responsible for the essential difference between low- and increase-phase voles. Although I disagree with its interpretation, this was a first-class experiment of the sort I had long hoped someone would carry out.

Figure 12.3. Number of young voles weaned by wild females (W) from low and increase phases and their descendants ($F_{1,2,3}$). From Mihok and Boonstra (1992: Table 1).

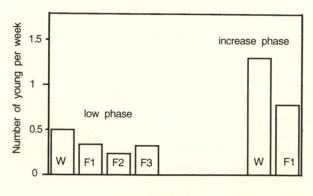

Generation

In a current manuscript the indefatigable Rudy and yet other colleague[15] have again exalted the role of maternal inheritance, this time in lemmings, and downgraded the relevance of any additive genetic component, even in behavior. In theory at least, motherly and grandmotherly effects can produce the lag required to explain the decline and low phases in some cycles. I doubt, however, whether such effects are a plausible explanation for long periods of decline and low numbers; but regardless of the final decision about the roles of maternal and genetic effects, it's always gratifying to see one's ideas taken seriously, if only to see them rejected or modified.[16] "A clash of doctrines is not a disaster—it is an opportunity." [17] Before I "collapse in deepest humiliation" [18] however, I await the outcome of further comparisons between low-phase, low-body-weight animals and increase-phase, high-body-weight animals. As I know from sad experience, it's often impossible to study cohorts of such animals at the same time, and in this latest study the only available animals were from an increasing population. Conclusions were therefore based on the study of one phase only, a defect I regard as fatal.

Thus, the evidence from John Clarke's study supports the idea that natural selection may act fast enough to play the part I claim for it; the evidence from Rudy's studies does the opposite. A third study, on larch budmoths, does both.[19] On the one hand, it supports the idea that selection has a role, since the morphs at the start of recovery mate assortatively and are selected against during the 4–5-year decline.[20] On the other hand, poor-quality food is said to be a necessary condition for the decline. As pointed out elsewhere, however, declines can occur when there has been little or no defoliation.

12.2 Alternative Methodologies

The "result" of an experiment is never the *totality* of observables; the result of an experiment is almost always the difference between at least two sets of observables.

Medawar[21]

Interest in cycles died out at Oxford after 1961, but an immense amount of work has since been done in North America (Figure 12.4) and Fennoscandia,[22] so I can hardly close this book without commenting on some of the studies most clearly related to my own[23] and without making suggestions for the future. As some workers believe that "hypotheses on intrinsic regulation of populations have flourished and collapsed," [24] I must try to counter this agnosticism. Objections to such hypotheses are of two kinds: that they are based on faulty methodology and that evidence for alternative hypotheses is more plausible than any so far produced for self-regulation (especially those involving natural selection). First let us consider the methodology, that is, the eliminative methods I hoped would prevent research from continuing down the same blind alleys. Yet according to Hilborn and Stearns, these very meth-

Figure 12.4. Some of the workers who have wrestled with the problem of what determines population density of small mammals. From left to right: Michael Johnson, Rudy Boonstra, Stan Boutin (kneeling), Rick Ostfeld, Dennis Chitty, Xavier Lambin, Charley Krebs, Michael Gaines, Brent Danielson. Photographed by Jack Millar at the Fifth International Theriological Congress in Rome in 1989.

ods have "retarded the analysis of population cycles in small rodents and of life-history traits." [25]

Hilborn and Stearns claim that the study of population regulation in small mammals "has been dominated by single-cause thinking" and that "if we accept on faith that every effect has a single necessary cause, we will fail to test the multi-factor hypothesis, and for this pay a price." For this deplorable state of affairs they place the blame squarely on me. Unjustly, however, for the authors themselves quote me as stating that my type of explanation involves "two or more factors in combination, including at least one necessary and sufficient condition." I should, perhaps, have put 'at least' in bold type; but 'two or more factors in combination' seems clear enough and would conform to one of the authors' definition of a multifactorial hypothesis— were it not for the word 'sufficient.' Unfortunately, the authors have muddied the waters by substituting 'sufficient' for the original word 'specific' and so have unfairly cast me out into outer darkness. Hilborn and Stearns continue:

> The essence of this [eliminative] approach is the search for *necessary* causes. It recognizes that disease, food, predation, and weather undoubtedly influence small mammal cycles, but rejects them as necessary causal agents.

Not so! Far from being rejected as causal agents, these factors are included among those necessary for the occurrence of cycles. What is rejected as unnecessary is a density-dependent *change* in these factors. This idea may be wrong, but should at least be attacked on its merits. In fact, I agree with those who "have argued that small mammal cycles are caused by an interaction of food, predation and social behavior," except that I would add to the list. Hilborn and Stearns are correct, however, in stating that "once a factor has been rejected as not a necessary cause of the decline, this factor is neither controlled nor measured in subsequent experiments." There are good reasons for this decision, namely, that natural events are affected by so many uncontrollable variables that instead of attempting the impossible job of studying them all, we restrict ourselves to studying those that differ consistently between a phenomenon and its control, or what passes for a control in population ecology. I have previously quoted Medawar and Born on the subject of comparative methods; but this methodology is of such paramount importance and so little appreciated in population ecology that I have quoted above yet another epigraph to the same effect. No apology is needed for preferring experimental to descriptive methods, nor for applying Occam's Razor and starting with the simplest explanation for *differences* between sets of data. With this approach, the starting point may indeed be the assumption of a single factor difference between a phenomenon and its control, regardless of the complexity of the phenomenon itself.

Hilborn and Stearns are not alone in failing to understand that 'necessary' means what it says—that and nothing more. It does not imply that cycles conform to "a stereotypic pattern without any geographic variation."[26] It does, however, imply that features common to all cycles have one or more common antecedents. Newton may have been wrong in stating that "to the same natural effects we must, as far as possible, assign the same causes," but should be given the benefit of the doubt.

The strongest advocates of a multifactorial approach is Lidicker,[27] who states that, "At least eight key factors (four extrinsic and four intrinsic) are required to explain the multi-annual cycles in [the California vole]." In his opinion, "a multi-factorial perspective is absolutely essential if we are going to understand microtine demography" and he warns against "the following logical pitfall: if A does not explain X and B does not explain X, then it follows that A and B can be omitted from any further explanation of X. This is the fallacy of multiple causation, because A plus B can still explain X . . . and is a trap that frequently ensnares microtine biologists." (If neither A nor B is *sufficient* for X, then A plus B may indeed be sufficient for X, but not if neither is *necessary* for X — which is all I claim for starvation, predation, and cyclic declines. Lidicker's misunderstanding comes from ambiguity in the use of 'explain.'[28])

In a "multi-factorial perspective" extrinsic and intrinsic factors are seen as 'inextricably interactive' and given that "the most influential ones can and do vary seasonally and with density, we are close to a full understanding of the density regulating machinery in the California vole." Such models "are of course testable;" for example, they predict that "complex processes like vole cycles will not be explicable in terms of single factors," and "demographic patterns often will vary spatially and temporally." While agreeing that every *phenomenon* is multifactorial, I suggest that this very complexity makes multifactorial *hypotheses* so easy to defend that they are irrefutable. Lidicker concludes as follows:

> If simple explanations for a phenomenon fail, complex ones may be closer to reality. In this particular case, a multifactorial model is supported by the evidence and by good sense. I am confident that it can guide us toward fruitful investigations and ultimately to powerful insights into the population dynamics of voles.

So I agree with another school of thought that while a multifactorial perspective 'seems reasonable,' it seems better still to examine 'simpler and more tractable hypotheses.'[29] Hanski et al. refer as follows to their ideas about specialist and generalist predators: "We suggest that any other hypothesis about small rodent cyclicity would also gain in credibility if it provides an uncomplicated explanation of these grand patterns in small rodent population dynamics."

Providing uncomplicated explanations is not, however, the way to make them credible, which they become if and only if they survive genuine attempts to refute them. While hypotheses must therefore be stated in the way best suited to encourage such attempts, that is, they must be tractable, authors must avoid the mistake of equating simplicity with familiarity. Everyone is familiar with the effects of predation, but it's systematic simplicity we are looking for, which means that hypotheses involving predation are suspect if they have to be protected *ad hoc* from instances little affected by predation. It would be better if future studies were planned without the reigning preconception that predators are necessary for the occurrence of cycles.

12.3 Lemmings in Alaska

Irrefutability is not a virtue of a theory (as people often think) but a vice.

Popper[30]

A nice adaptation of conditions will make almost any hypothesis agree with the phenomena. This will please the imagination but does not advance our knowledge.

Black[31]

Frank Pitelka and I first met at Cold Spring Harbor, where he voiced his doubts whether cycles in the arctic "are in basic principle the same as those of mid-latitudes."[32] He continued to be sceptical during his 1957–1958 sabbatical year at the Bureau and during 1962–1964, when he sponsored Charley Krebs during his tenure of a Miller Fellowship at Berkeley. He is still sceptical.

Frank Pitelka is another advocate of an alternative approach to solving the cycle problem: "regulation of numbers in a cyclic lemming population is really part of a nexus of regulatory interactions in tundra as an ecological system."[33] During visits to Point Barrow, Alaska, between late May and early September Frank obtained a long record of fluctuations in the numbers of lemmings (Figure 12.5). Before the start of this snap-trapping, brown lemmings were known to have had peaks in 1946, 1949, and 1953. Later fluctuations are discussed with reference to the 'nutrient-recovery hypothesis.'

According to this hypothesis, "the cycle is a result of interaction between herbivore and vegetation mediated by factors of nutrient recovery and availability in the soil."[34] The quality of the vegetation, after being spoiled in the peak year, gradually improves over the next "two or three years of no grazing"[35] and the return of nutrients from excreta produced in the peak year. The regular 4-year cycle predicted by this hypothesis was modified, however, by a series of events that raised "questions as to whether the nutrient-recovery hypothesis . . . adequately explains cyclic fluctuations in the brown lemming."[36] Three factors had apparently disrupted expectations from this hypothesis: (a) unusual weather may have deflected the normal course of some cycles but was unlikely to have been responsible for all the irregularities of 1965–1972; (b) immigration may have come from inland areas unaffected by the severe weather on the coast during 1969–1970; and (c) predation was disorganized during 1965–1972 and failed to play its part in setting the stage for the "regularity, strong amplitude, and relative timing of normal microtine cycles in the Arctic." As a rule, predators "slam the prey down and set a time Zero for the next cycle;" but sometimes "this co-adapted system fails." De-

Figure 12.5. Number of lemmings caught in June at Point Barrow, Alaska. Collared lemmings were seldom caught before 1968. Numbers refer to events, described in the notes, that differed from expectations from the nutrient-recovery hypothesis. For June 1958 I have added 1 to a catch of zero. From Pitelka (1973: Table 1, which also includes data for July and August).

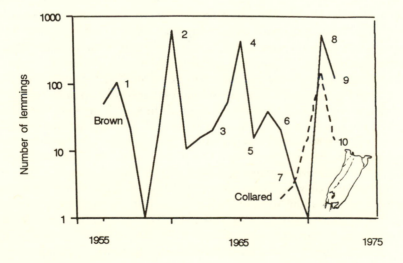

tails of these disruptive events are given in the notes,[37] though it's not clear how some of them were obtained after the author's fall migration to the south.

The attraction of multifactorial explanations is that they can explain almost anything. Their disadvantage is that, through permitting a wide range of events, they are correspondingly difficult to falsify. I agree with these explanations only in doubting the adequacy of the nutrient-recovery hypothesis. As "lemmings attain the highest body weights in the peak year"[38]—"the time of overgrazing"[39]—the explanation for their fluctuations is probably the same as for other microtines, contrary to Frank's original doubts.

Future workers will have to decide whether to continue the so-far uncorroborated faith in the efficacy of density-dependent changes in food supplies and predation, or to accept the advice, given over half a century ago, to look in other directions. I believe this advice is sound and that if more of my contemporaries had accepted it they might already have written the concluding chapters to this over-long detective story.

12.4 A Last Hurrah

There are many things that hinder sure knowledge, the obscurity
of the subject and the shortness of human life.

Protagoras

Thirty-five years ago, when I disappeared into teaching, I expected to see a mad rush of workers testing the idea that natural selection was a necessary component of cycles. One such paper appeared in 1968, independently of anything I had written,[40] and encouraged Charley Krebs and his students to study population genetics while they were together at Indiana in 1966–1970 and after they went their separate ways. Results were encouraging at first in showing that rapid shifts in gene frequency were sometimes associated with rapid declines in numbers,[41] but attempts to select for and against certain genotypes were largely unsuccessful. A devastating blow to these attempts was the discovery that isolating a population, even in enclosures as large as 0.8 ha, destroyed the very phenomenon being studied.[42] Instead of leveling out at moderate peak densities, as in unfenced control populations, numbers continued to grow until the animals reached three or four times their natural densities, ran out of food, and starved to death.[43] This artifact, which became known as the 'Krebs fence effect,' made it impossible to establish realistic populations of different genotypes. Other difficulties damped ecologists' enthusiasm for becoming amateur geneticists. In the Vancouver area, for example, *populations* of voles fluctuated irregularly, and the *species* was stated, wrongly in my opinion, to be noncyclic.[44] As well, the animals proved to be difficult to breed in small cages and lacked the genetic variation expected from the Indiana studies.[45]

These difficulties took the glory off my second coming, for under the stimulus of Charley's return to UBC in 1970, I decided to spend 1972 doing a so-called crucial experiment. By then there was plenty of evidence that voles are capable of regulating their own numbers but none that natural selection is a necessary part of the process. The simplest way of testing this hypothesis, it seemed to me, would be to find two out-of-phase populations, breed from their pregnant females, look for differences in the behavior of their progeny, and see how they reacted to one another. Alas for simple ideas: every female resorbed or aborted her fetuses. It turned out that, with this species, the only way of guaranteeing success with pregnant females from the wild was to wait until they were within 3–4 days of parturition. But by that time I no longer had out-of-phase populations at my disposal.

Anyone prepared to try this approach should be aware that a lab environment that suits animals from (say) the increase phase may be unsatisfactory for animals from other phases. This is indeed likely if social and reproductive behavior differ throughout the cycle. Particular difficulties may be found in establishing a line of animals at the start of the low phase, which includes individuals that sometimes breed poorly and make few or no gains in numbers or body weight. At some unpredictable point, however, the phase ends with animals that are indistinguishable from those in the increase phase, which may explain why samples from the field are a mixture of animals that grow at different rates in the laboratory. Were it not for the difficulty of getting enough of them, studying low-phase animals would be the simplest way of discovering qualitative differences between those that end the decline and those that start the recovery.

Instead of bringing in pregnant females, I next tried to build up domesticated stocks from animals mated in the lab in successive years of a cycle. Alas, again, for simple ideas: few matings were successful in small cages;[46] also, the only population I had time to study lacked the clear-cut phase contrasts that would have enabled me to build up pure lines. In large outdoor enclosures, however (mine were 21 sq m), it was easy to breed voles in colonies, which provided animals for setting up artificial populations in the wild and making observations in the lab. The results of the field studies have been published;[47] those from the lab were chiefly important in destroying my belief in the docility of the large, fast-growing animals typical of the increase and peak phase. I also learned much else about aggressive behavior but little that had not been observed in other species of mice and voles.

Here I shall describe only one of these confirming observations, namely, the formation of a class of subordinate animals like those described by John Clarke. One of my colonies lived indoors under sheets of plate glass (2 ft x 1.5 ft, resting on a maze of 2 x 4 s) spaced 12 inches apart in a room of 23 sq m. Otherwise, the set-up was similar to that in John Clarke's outdoor snake pit. As the animals made roofless runways in their litter, it was easy to record movements and interactions. In the company of Janine Caira, an honors zoology student who has since become a distinguished defector to parasitology, I spent many hours recording the spacing behavior of these animals and made a census each week by catching the animals and keeping them in cans until all had been weighed. After being released at the weighing station, most animals dispersed amicably; but after a while we noticed an increase in fighting, which continued until those attacked had returned to a 'ghetto' under one of the glass plates. The ghetto consisted of a cluster of six or more animals that remained almost totally inactive during our watches and must have run the gauntlet when they did venture abroad in search of food. Occasionally, a vole from the rest of the colony would enter and display aggressively but not otherwise molest these evidently subordinate animals. (As soon as we had figured out what was going on we returned these second-class citizens directly to the ghetto.) On other occasions subordinates escaped persecution by climbing up the sides of the wire-mesh wall mentioned below.

In the wild, such animals might have dispersed (only to be killed on neighboring territories, perhaps); or if they had stayed might have contributed to the problem of nonrandom sampling. Their behavior was consistent with the long-held suspicion that the animals least easy to trap are subordinates that are wary of getting caught in traps visited by dominants.[48]

Three other results are worth putting on record. (1) We discovered that strangers put into a colony were not the only individuals to suffer from aggressive behavior. During the ensuing fighting some of the residents attacked one another, much as conflicts between coveys of partridges led to fights within them. Similar disruption may be what happens in the field owing to immigration of strange voles, especially in spring. Relatives and nonrelatives may suffer alike, which would confirm Hamilton's belief that spiteful behavior may reduce the mean fitness of a population.[49] (2) Spring movements are probably an effect of increasing daylength. We tried to mimic the change from winter to spring by first putting a colony on short daylength and then switching to long daylength. This is probably the best way of carrying out a laboratory study of the "spring decline," as the effects on growth were striking (Figure 12.6); but we failed to invent techniques for mimicking the dispersal that probably accompanies it in nature. (3) By having a fenced-off population down one side of the room, we learned to recognize which animals displayed against their neighbors. Dominant males and pregnant females used to make periodic patrols along the borders of the wire mesh enclosure, displaying violently against the occupants.

Apart from giving these notes, I shall not describe what else I did with my time. I should have realized I had a tiger by the tail and stuck to my teaching and supervision. Although I failed to get to first base myself, I did at least provide facilities and animals that students could quickly get to work on. Supervisors sometimes have their uses. Besides former students whose help I have already acknowledged, several others took up related aspects of the problem of the regulation of numbers. Mary Britton, Wren Green, Mike Healey, Art Lance, Richard Sadleir, and Cheryl Webb made valuable contributions before entering other fields. In conclusion, I thank these and other students who have made their supervisor's career such an enjoyable one.

12.5 Quo Vadis?

Research is intellectual adventure; and the universities should be
homes of adventure shared in common by young and old.

Whitehead[50]

Medawar's *Advice to a Young Scientist* can be read by people of all ages
and scientific interests.[51]The present section, by contrast, is narrowly directed
towards supervisors and students in search of a topic for a MSc or PhD thesis
dealing with cycles. I will not refer to the many promising lines of enquiry
I've had nothing to do with, but will stick to a few I've been involved with
either directly or indirectly through correspondence or personal discussion.[52]

Before getting down to specifics, students should recognize that population
ecology is in disarray and that lack of consensus is largely due to failure to
agree on standards of scientific evidence. A critical review of the past may
encourage them to substitute experimental for descriptive methods. In as-
serting the value of 'strong inference,' Platt might have been referring to the
population ecology of small mammals when he wrote[53] "A failure to agree
for 30 years is public advertisement of a failure to disprove," though I would
substitute '54' for '30,' change 'disprove' to 'refute,' and add that the result
has been a lack of incentive to look for the cyclic equivalent of the Rosetta
stone. (It's 54 years since Charles Elton said, in effect, that it's a waste of
time trying to explain vole cycles in terms of starvation, predation, or specific diseases.)

Population ecology is an exceedingly important discipline, not only for
what it adds to our understanding of the natural world, but for what it can
do to protect endangered species, conserve our dwindling stocks of fish and
other commercial animals, and protect us from insect and other pests. It will
fail to meet these objectives unless it subscribes to the highest standards of

Figure 12.6. Effect on body weight of change from short to long daylength. Two
months on 8L:16D were followed by five weeks on 16L:8D. Townsend's voles.

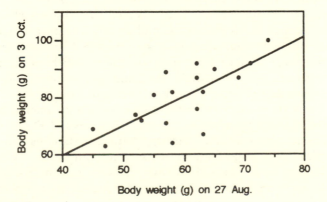

scientific evidence. My harping on the need for properly controlled experiments will, I hope, speed up the present drift in this direction and show our colleagues in the traditional disciplines that there's hope for us yet. Eventually, it may also persuade government departments to refrain from uncontrolled predator reduction policies, though this is a less sanguine hope.

The problem of most general relevance to population ecology is, I believe, to discover the differences between cyclic and noncyclic populations,[54] for there's obviously an element present in the one that's absent from or less pronounced in the other. This undiscovered element may be what's needed for making the falsifiable predictions so sadly lacking. The practical difficulty is to find both kinds of population living fairly close toghether. For the farther they are apart, the greater the number of irrelevant differences between them, and the greater the logistical problems of making comparable long-term studies on both. In addition, one would have to compare populations in recognizable phases of a cycle with populations that are unambiguously noncyclic. In spite of the implied but doubtful assumption that so-called noncyclic populations are fundamentally alike, several workers have gone ahead anyway and established laboratory stocks that do indeed differ according to their origins.[55] In principle, I recommend work such as that done to compare cyclic populations from the north of Sweden with supposedly noncyclic populations from the south (though some authors fail to state the phase of the cycle from which they obtained their northern samples).

The most ambitious program was carried out at the University of Umeå, where geneticists looked for "consistent changes in genotype frequencies during populations cycles" and compared differences between cycling and noncycling populations.[56] Their program got off to a promising start, especially in discovering differences, some of them heritable, in social behavior, activity,[57] and response to photoperiod between voles from north and south Sweden. The difficulties of this type of work soon became obvious, which may explain why most ecologists shy away from it. The program covered one cycle only before grinding to a halt as one of the authors, like many a doctoral student, took off for a job elsewhere.

A consortium of no less than 10 other Swedish scientists cast a jaundiced eye on this type of approach.[58] They claimed that distinguishing between cyclic and noncyclic populations "has had a strong hampering effect on the development of small mammal population research." They are not the first to point out "that the underlying basic mechanisms behind the population dynamics of small mammals may be the same for all populations" (they ignored my 1960 and 1977 statements to this effect), and claim that the differences are "of marginal interest for the understanding of small mammal ecology in particular, and populations dynamics in general." Fortunately, other people took the opposite view, namely, that understanding differences (as well as similarities) will go far towards solving both particular and general problems. I regret having missed my own chances to discover differences between cyclic populations of field voles and noncyclic populations of bank voles living in the same habitats. Such as analysis would have avoided the possibly confounding effect of geographical and habitat differences that cast at least a small shadow of doubt on conclusions from the Swedish work.

A more practicable approach to finding qualitative differences would be to continue the one I misinterpreted back in the 1930s, namely, trying to find differences associated with peak and decline, where the decline is the phenomenon to be explained and the peak is the control. Later on it seemed better to regard the increase phase as the control for all other phases, especially the early phase of low numbers among cyclic populations and the commoner stationary state of other populations.

Figure 12.7 shows the problem as revealed by two studies 40 years apart but here supposed to describe a single population going from a cyclic to a stationary state, that is, a state of seasonal fluctuations only.[59] After a minor peak in numbers in the spring of year 1 (1972 in real life) a population on Westham Island declined to low numbers in springs 2 and 3 in spite of good recruitment in the previous autumns. It then increased slightly more than in these 2 years and now remained at a high peak the next spring.[60] It declined in year 5. We need to know why overwinter survival in years 2 and 3 was not equal to that in year 4 and why numbers crashed in year 5 but not in years 4 and 6-9. The attraction of this approach is that with luck, and in spite of confounding differences between years, contrasts are so well marked that they show up as differences in behavior in the field and among animals taken into the lab.

As we already know, the fluctuations at Corris were not preceded by unregulated increase from a low phase. We therefore had no data with which

Figure 12.7. Number of voles in autumn (●) and spring (○) in a cyclic and non-cyclic population. For Westham Island the data are the minimum number alive on 0.8 ha (Krebs 1979: Fig. 2). For Corris the data are from an index of abundance (see Figure 2.2). ? = problems discussed in text.

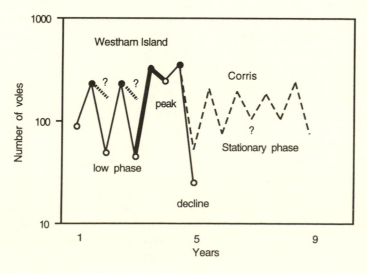

to contrast its regulated state. In noncyclic populations such contrasts are seen only by luck during recovery from natural catastrophes, so need to be produced experimentally.

Luck, unfortunately, is seldom on the side of the field worker. Populations may fluctuate in unrecognizable ways, may suddenly cease to cycle, or may get stuck at low densities. So, as time is of the essence for graduate students, a fundamental principle is to do a variety of pilot tests. These should take up just enough time for the student to judge the most promising lines of attack and to jettison the rest without further wasted effort. The importance of making the right initial decisions cannot be overemphasized: if a job is not worth doing, it's not worth doing well. I recommend that students take 3 to 6 months at this stage and be given complete freedom to choose a project on which they are prepared to stand or fall. They should never be used as slaves laboring for the greater glory and productivity of their supervisor. Treated this way they are unlikely to be much use when on their own.

One of the difficulties of starting a study on cycles is to recognize the population phase so that one can get on with experimental work instead of merely adding one more to the long list of descriptive but otherwise barren studies. With a population in the increase phase, a useful procedure would be to try to keep it there by heavy cropping. Attempts to do so have been frustrated by the surprisingly high rate at which the density of voles and bank voles is restored by immigration. Baltensweiler had the same trouble with larch budmoths.[61] Until we know how to prevent populations from declining, we cannot substantiate our claims to understand why they do or do not do so.

So far, no one has tried to find out experimentally whether populations are synchronized by weather. Chance may favor the mind prepared to recognize populations that are out of phase and try to understand how they come back into it (if they oblige or frustrate one by doing so); but it would be better to devise procedures for getting them into and out of phase. The rival point of view — that predators are responsible for the synchrony —is, I think, untenable. I see no evidence that cycles have the biological properties one would expect from predation. Losses due to predators are inadmissible evidence without experimental evidence that declines and low phases are absent when predators are also absent. No one has yet done more than modify rates of decline.

Other studies need a stock of lab animals or the possibility of finding enough wild animals to create a breeding stock from scratch. The latter approach, as I know to my cost, is fraught with danger; for if the animals refuse to breed, the thesis will leave much to be desired. For those willing to take a chance, however, there's gold to be mined from understanding the factors that inhibit or stimulate reproduction and cause lab stocks to diverge from their wild forebears. Let us assume, however, that a kindly supervisor puts his breeding colony at the student's disposal. Several projects are now possible. I myself learned most about vole behavior from watching animals living under plate glass on the floor of one of the rooms at the Animal Care Centre at

UBC. It was sometimes difficult, though, to identify animals at the far end of the room, and for this reason I found a gallery system more satisfactory. Also, the linear design makes it easier to record movements and interactions.[62] Varying daylength is one of the manipulations likely to be profitable, not only because of its physiological effects, but because changing from short to long daylength is a possible way of producing the lab equivalent of a spring decline. Provision should be made for recognizing dispersers, perhaps by giving them a fence to climb up to get out of harm's way.

All conclusions from lab studies must be tested in the field. Direct observation of a species that lives under heavy cover may be difficult, but perhaps something may be learned from attracting individuals to baits, as we did with rats and as Kikkawa did with bank voles, or by staging encounters within a resident's home range. Wolff and his colleagues discovered that at high population densities resident deer mice nearly always won through aggression, but that at low population densities "individuals of the same sex maintained exclusive home ranges by mutual avoidance."[63]

If natural populations are scarce, the best substitutes are fenced populations. Crying aloud for further study is the 'Krebs fence effect' (the failure of some enclosed vole populations to regulate their numbers). Though not everyone will have the resources to study them on the grand scale shown in Figure 12.8,[64] it's possible to create smaller fenced populations that also behave like their unfenced controls. Much has still to be learned from this ap-

Figure 12.8. Bob Tamarin (on right) and his intellectual grandfather outside the 0.8-ha Grid K of Tamarin et al. (1984). The size of the enclosure and number of replicates have increased since the days of the 'snake pit' (see Figure 6.3). Photo in October 1990 by Sherry Kendall.

proach,[65] but it may be even more profitable to study the aberrant behavior of totally enclosed populations. They are spectacular examples of the failure of self-regulation (comparable to that of house mice in Australia)[66] and so provide another type of control for populations that are self-regulated. Voles in enclosures probably survive in the form of amicable extended families with strict territorial boundaries. We can guess that lack of dispersal is the main reason why numbers build up within these families and that it has two consequences: (1) that offspring live unharmed as subordinates instead of dispersing and getting killed, and therefore (2) that they don't disrupt the social organization of their neighbors. It would be instructive to see what goes on at these territorial boundaries, perhaps by seeing if they are maintained under the cover of plate glass instead of invisibly under natural cover. 'Spacing behavior' is still a black box. One can also learn much from radio- and other forms of tracking.[67]

Some workers are still hostile to the idea that populations of small mammals are able to regulate their own breeding density and thus set limits to their annual rates of increase. The three most diagnostic features of this process are large and sudden drops in numbers, especially in spring, poor recruitment, and delayed recovery from scarcity. These observations are not consistently explained by the action of extrinsic mortality factors. Some workers fail to recognize that they have these problems. Failure to recognize the suddenness of a spring decline is sometimes due to sampling less often than every week or so and to plotting the data arithmetically. The significance of poor recruitment and persistence of the low phase may also be unrecognized. Through lack of controls, losses over winter may be dismissed as mere seasonal effects, whereas they seem to be due to the elimination of surplus animals produced in fair numbers in the previous year. Numbers in autumn are thus unreliable indicators of self-regulation.

The perceptive reader will have noticed my belief that mathematical modelers have done more harm than good through possessing a medieval faith in the power of logic to unravel secrets of the natural world.[68] Whitehead has warned against this delusion.[69] The danger is that, instead of being "used to challenge existing formulations," models may "conform our biases and support incorrect intuitions."[70] Some of my best friends are modelers, and those who are putting ideas into refutable form are obviously doing the right thing.[71] My concern is for the inhibitory effect of certain disk-bound elder statesmen on the independence of thought of younger scientists. Nevertheless, if theorists were to be disciplined by the untidy observations of their ecological colleagues, or were themselves to test their own hypotheses, they might be more use than they have been so far. Time spent watching a screen might be better employed watching the behavior of real animals.

Leonardo da Vinci said it all when he wrote that "those sciences are vain and full of errors which are not born from experiment. . . and which do not end with one clear experiment."[72] Five centuries later, however, we can no longer endorse the omitted phrase—"the mother of all certainty." For the only certainty, according to Eddington,[73] is that your pet theory will be wrong if it's against the second law of thermodynamics. This is not a line of argument I recommend to students defending a thesis, but they should stick to their guns for awhile, for not only may theory be wrong, but "experimentalists do bungle things sometimes." In spite of their bungling, however, they and not the theorists are the final arbiters and as such are morally obliged to bungle things as seldom as possible.

As one who believes that cycles are merely a special instance of a general problem, I suggest that students feel free to choose some other organism than a small mammal. If self-regulation is common throughout the animal kingdom, many species would make better experimental animals; and if natural selection is involved, the principles should apply across the taxonomic board. The differences between voles, lemmings, snowshoe hares, game birds, larch budmoths, spruce budworms, and other species suggest that something more general than their social systems must be responsible for similarities in their population dynamics. One factor common to all fluctuating populations is systematic change in the number and kind of neighbors surrounding each individual. If, as I suspect, individuals are differentially sensitive to this environment, different phenotypes will predominate at different stages of population growth. We know that gene frequencies can change rapidly during a decline,[74] and that natural selection can act a good deal faster than we used to think. So my guess is that natural selection is the agent responsible for *essential* changes in the quality of cyclic populations. This view

> is appealing because it follows so well from Darwinian theory, i.e., nature selects those organisms that are best suited for a set of environmental conditions. When conditions change, as happens when population density increases, the frequency of the favored genotype, behavioral type, morphotype, or other types may also change.[75]

An alternative explanation for qualitative change is that it is a maternal effect. Long-standing evidence of cytoplasmic inheritance in locusts[76] suggests that one must keep an open mind on these issues and that whichever explanations one starts out with is likely to be rewarding. Settling these questions will probably require analysis of DNA polymorphisms, which will give precise kinship data.[77] Early results have been promising.[78]

So far, the best study of qualitative differences is that carried out on red grouse.[79] For the whole of an 8-year cycle, young grouse were bred in captivity from the eggs of wild parents and were themselves bred from in the following year. By comparing dominance and aggression within and between year classes, including fathers versus sons, the authors concluded that aggression and dominance were both heritable and that "Changes in spacing behaviour between the increase and decline phases of the population fluctuation [were probably] the main cause of the changes in the rate of emigration and in consequent densities. What caused these suggested changes in behaviour remains unclear."

According to the authors, these results refute one version of my polymorphic behavior hypothesis, though we seem to agree on three important issues, namely, that spacing behavior, natural selection, and average quality are all associated with cyclic phenomena. An alternative hypotheses is that, owing to falling rates of recruitment as numbers build up, there's an increase in old:young ratio. This in turn implies that neighbors are less closely ralated and therefore more antagonistic than those earlier in the cycle.[80] Predictions consistent with this idea have been confirmed but are also consistent with other ideas, from which it has not yet been separated experimentally. Testing the idea on voles showed that groups of related animals survived better than other groups, but that these differences in survival made little difference to the course of the cycle.[81]

Any system common to all cycles will be modified by local circumstances, some of which will account for differences between populations in rates of change and average density. These extrinsic factors need to be studied, but seem unlikely to explain the similarities among cyclic populations.

It may seem strange that after more than 70 years' work so many problems remain unsolved. Perhaps, as I have suggested throughout this book, they owe their long life to ecologists' reluctance to use eliminative methods, which would have compelled them to look in new directions. It will be better to end on a positive note, however, by pointing out that success of a research program depends on how long it continues to generate new lines of enquiry.[82] By this criterion, studying cycles has been a success. We may not have solved the problem, but we've certainly learned a lot since 1923, when Charles Elton set the ball rolling. Like the Mississippi, it just keeps rolling along. No scientist could ask for a finer memorial.

Notes

'Anything that interrupts the flow of a narrative is an abomination of desolation. Readers should therefore ignore all notes until later, when they can read them as a supplementary account.

Preface

1. Schrödinger (1954). For the rest of this passage, see Section 9.2 note 8.

2. Popper (1963); also Wikander (1985): "Testability of an idea offered or regarded as a *scientific* (as opposed to, say, a metaphysical) hypothesis is absolutely indispensable to both the process and progress of science, and indeed is part of what is intended by the use of the term 'science.' "

3. Simberloff (1983) says: "The search for confirmatory evidence is easier than the search for falsification and is very seductive. . . . confirmatory evidence per se is not very compelling since one can always find some."

4. Thomas Henry Huxley (1902, CE 8:240): ". . . the great tragedy of science—a beautiful hypothesis slain by an ugly fact."

5. Scientists and nonscientists should read some of the same books; but in trying to kill two birds with one stone I here run the risk of merely wounding both.

6. Avoiding the first person singular is wrongly supposed to make scientific writing more objective. According to Hull (1988: 199), "Whatever scientists are learning in universities, it is not how to write clearly and effectively." Once acquired, poor writing habits persist, as students metamorphose into professors, deans, editors, and members of learned societies, with catastrophic effects on standards of scientific prose. "There is no form of prose more difficult to understand and more tedious to read than the average scientific paper" (Crick, 1994: xiii).

7. Bacon in Anderson (1960: 156).

8. Like other human pursuits science often departs from its ideals (Gardner, 1981; Bell, 1992; Taubes, 1993).

9. Crick (1988: 142)

10. Vogel (1992).

11. Crick in Judson (1979: 113).

12. Some critics fail to understand that a necessary condition for a phenomenon is never its sole condition and cannot be said to cause or, worse still, to 'drive' it. The use of the word 'cause' should be restricted to factors that explain consistent differences between a phenomenon and its control. Critics with other views about the regulation of numbers automatically reject any feed-back mechanism that depends on behavior rather than density.

13. Beveridge (1950: 89).
14. Whitehead (1942: 129).

Acknowledgments

1. Judson (1979: 357). It was then that Monod "discovered what science was about."
2. Beveridge (1950).
3. Whitehead (1953: 75).
4. Macfadyen (1992) describes the wide range of Charles Elton's interests; his work on cycles was only one of many contributions beyond the scope of the present book. See also Berry (1991).
5. *Peter Crowcroft obtained his D.Phil. in 1954 for work on shrews. He later worked on house mice and published his findings first in conventional form and later in a light-hearted book* Mice All Over *(1966). Its style, I wrote, owes "more to Rabelais than to the* Journal of Comparative and Physiological Psychology" *and suggested that "those who might feel embarrassed about enjoying this kind of literature should ask the publishers to mail them* Mice All Over *in a plain envelope." I wish more research was written up in this lively style. Peter's 1991 book,* Elton's Ecologists. . . . *was written out of gratitude for all he owed to Charles Elton. This book was reviewed favorably by C.J.Krebs (1992a), with reservations by J.R. Krebs (1992), and unfavorably by Hagen (1992). The latter deplored the author's failure to place the "research within the broader context of twentieth-century ecology," to deal with relationships between Elton, Lack, and Tinbergen, to investigate more thoroughly the demise of the Bureau, and to have painted the rest of the Bureau as rather drab, uninteresting characters. As one of these drab characters, I consider this review unfair. The book does brilliantly what it sets out to do, namely, capture the spirit of the Bureau. It may not be the kind a historian of biology would have written, but it's a rich source of material for anyone wishing to write such a book and to understand what went on behind the published records.*
6. Niko Tinbergen later shared the Nobel Prize with two other ethologists. He supervised a lively group of students working in the Department of Zoology at Oxford on a variety of problems of animal behavior. According to Medawar (1984a:200) ethology was in the fortunate position of not being "crabbed and confined by the doctrinal tyranny of any pre-existing explanatory system." Population ecology has been less fortunate, though John Krebs (*in lit.*) claims Medawar's view was too rosy.
7. Szent-Györgyi (1964).
8. From Crick (1988: 95) we learn that Andrew Szent-Györgyi was one of the hoaxers who sent out invitations, forged in 'Joe' Gamow's characteristic handwriting, to a "whiskey, twisty RNA Party" in the latter's cottage. Jim Watson was one of the other hoaxers. Although he seldom played practical jokes, "his mentor, Max Delbrück, was notorious for them."
9. Elton (1942).

Chapter 1 Introduction

Section 1.1

1. Feynman (1966) in his Nobel Lecture. See Mackay (1991).
2. Vogel (1992). The epigraphs are designed to place the present study in a broad

scientific context. They are also intended to help graduate students, most of whom are too busy collecting new knowledge to brush up on old knowledge about the history, methods, and philosophy of science. Yet if they are to become effective teachers they must do more than prattle endlessly about their specialty: they must show students the wood by means of the trees, not prevent them from seeing the wood because of the trees (Whitehead 1950: 10). I believe that all science students should at least dip into writers on the philosophy of science. I particularly recommend chapters 1–6 in *Science and the Modern World* (Whitehead 1953). Johnson (1947) extols Whitehead for "an unexpected brilliance of vivid expression . . . [and] a sparkling somewhat ironic humor." Medawar (1984b: xi), however, was nearly put off philosophy by Whitehead's "extreme length, leaden prose and general air of joyless learning . . ." You pays your money . . .

3. Whitehead (1953: 3).

4. One of the reviewers of my manuscript complained that it was "rather jumpy, with many very short sections and frequent shifts of topic." I agree; but the style is intentional, as I wish to bring out the contingent nature of discovery, a characteristic that is concealed in the apparently logical progression of events described in scientific papers. First, events were shaped by such uncontrollable variables as money, the interests of one's students, discussions over coffee, recent literature, other commitments, conferences attended (Bondi, 1962), and so on. Second, thoughts about these events cropped up at the time or up to 70 years later. Third, since this is a mainly chronological account, events need to be described as they occurred. I have therefore tried to make each section short, snappy, and more or less self-contained—the written equivalent of the sound-bite—so that readers can pick and choose and, with the help of the index, relate events to those before and after. This format entails some repetition, for which I apologize to those readers "who demand full value for money from every printed page" (Medawar, 1984a: 19).

5. "How odd it is that anyone should not see that all observation must be for or against some view if it is to be of any service!" (Darwin in Barlow, 1958: 161).

6. Qualitative differences were known in locusts, which change from a *solitaria* phase at the start of an outbreak through a *transiens* to a *gregaria* phase; but in the 1930s there were no intellectual bridges between those working on locusts and those working on other fluctuating populations. Nowadays both groups of workers have the common problem of trying to understand the roles of maternal inheritance and natural selection in effecting changes in quality. Krebs (1985) gives a good account of the locust problem.

7. Dunbar (1980). See replies by Holsinger (1981) and Wikander (1985).

8. McIntosh (1985).

9. Watkins (1964). If rewritten in a politically correct style this passage would lose its rhythm. Nevertheless, one cannot read this passage in 1996 without feeling embarrassed "by the lack in English of an epicene personal pronoun or possessive adjective" (Medawar, 1979: xiv). Rather than mangle the English language, however, I apologize for my own gender-neutral use of 'he' and 'his.'

Watkins comments further on the 'didactic dead-pan' style of most scientific writing: "if natural scientists . . . took to writing in a candid, uncensored, autobiographical way, setting out their ideas in their natural order . . . a gratuitous barrier to mutual comprehensibility would have faded away." Medawar (1986: 141) was also "one of those who feels quite strongly that writing, however solemn or formal, should sound like speech." Hence my use of contractions, which may irritate some readers, but for which Vogel (1992), a lucid and successful writer, is my role model.

10. See p. 208 for an example of a historian getting things wrong. Like many reviewers, he suffers from the chronic ailment of wanting a book written in his style instead of the author's.

11. Hull (1988: 305), who epitomizes the factors driving scientific research as *curiosity, credit, and checking*. According to Bernard, who may not perhaps, represent the opinions of all Frenchmen. "The joy of discovery is certainly the liveliest that the mind of man can ever feel" (Beveridge, 1950:140).

12. Senator Proxmire's annual award of the Golden Fleece went to the title that sounded most ridiculous.

Section 1.2

13. Popper (1963: 38).

14. Elton (1942: 215). Marsden (1964) quotes an equally incredible story of "untold thousands" of lemmings plunging into the waters of the Canadian arctic.

15. Myllymäki et al. (1962) quantified directional movements by counting the numbers crossing a 200-m wide stretch of water between Sweden and Finland. The majority of lemmings were bound for Finland. Like 'wetbacks' crossing illegally into the U.S.A., most moved under cover of darkness.

16. Koponen et al. (1961).

17. The following account appeared in the *Manchester Guardian Weekly*, May 29, 1983: "WALT DISNEY made snuff movies. You will remember the classic lemming scene in White Wilderness (1958). Brian Vallee, a Canadian TV producer, has spent five months researching the making of such touching displays of animal phenomena in Disney movies and has established that 1,000 of the little rodents were bought off Eskimo children and flown to Alberta.

When the wretched creatures refused to commit suicide on cue the Disney crew obligingly stepped in and threw them off the edge of a cliff into a river."

European lemmings would have been equally uncooperative.

18. Henttonen and Kaikusalo (1993).

19. H. Chitty (1950b) includes references to all previous reports.

20. We were unaware that Sutton (in Sutton and Hamilton, 1932) also claimed that lemmings bred throughout the winter. Sutton based his claim on observations on Southampton Island in 1929–1930, in contrast to earlier observations by Soper (1928) that lemmings did not breed in winter on Baffin Island. The apparent inconsistency was cleared up by Krebs (1964a), who discovered that winter breeding was uncommon except in the increase phase. This was probably the phase Sutton observed, as lemmings were so abundant that hunters had to kill 'hundreds' of them if they hoped to sleep nights.

21. Birkeland (1986). This account was kindly sent me by Tarald Seldal and Helge Monsen of the Museum of Zoology, Bergen. It was translated by them and Karen Hidden.

Section 1.3

22. Bacon in Anderson (1960:130).

23. Shelford (1943).

24. Elton (1942, Table 51).

25. For other advantages to using logarithms, see Williamson (1972).

26. See Section 2.5 note 36 for comments on predator-prey models.

27. Gross (1947).
28. Anderson (1939).
29. My next attempt to study lemmings was also frustrated.

In 1952 John Tener, of the Canadian Wildlife Service, was duly enrolled as my student but was unable to get more than 1 year's leave of absence. Although he sent over an advance guard of lemmings, he was able to work short term only on other projects (Leslie et al., 1955). My main contribution to his education, he told me recently, was my insisting that he join his college rowing club (St. Catherine's). Engaging in sports and living in college, or at least eating there, are an integral part of the Oxford system of education. John later earned his Ph.D. at the University of British Columbia, working at the other end of the zoological spectrum on muskox.

30. Potts et al. (1984), in discucssing fluctuations in red grouse, adopted (from other authors) the following terms: phase forgetting quasi-cycles; phase remembering quasi-cycles; exogenous resonant quasi-cycles; and perturbed limit cycles.

31. Shelford (1943,1945). Peak years in the Churchill area are given as 1929, 1933, 1936, 1940–1941, and 1943. Lemmings here were apparently out of synchrony with the 1937–1938 peak reported elsewhere (H. Chitty, 1950b). May (1981) has fitted a theoretical curve to Shelford's 1943 guestimates of population density, a somewhat easier task than trying to fit theoretical curves to the more reliable data that have since become available for this and other cyclic species.

Section 1.4

32. Reichenbach (1959: 5).
33. Batzli et al. (1980).
34. Peters (1991: 297); see also Whitehead (1942:138): "The preoccupation of science is . . . the search for simple statements which in their joint effect will express everything of interest . . . ."
35. Reichenbach (1959: 5).
36. Whitehead (1953: 64–72).
37. Bernard stresses the necessity of making comparative experiments "always and in spite of everything . . . I might quote a great many errors into which able experimenters have fallen by neglecting the precept about comparative experiment." (Green, 1957: 181–183).
38. Many of the factors common to a phenomenon and its control will be necessary for both. So while a difference between them may be due to a single factor, the phenomneonon itself will occur if and only if other necessary conditions are also satisified. Much controversy has arisen from failure to see that one can study necessary factors without claiming that they are sufficient. The misleading label ('single-factor hypothesis') sometimes pasted onto my own hypothesis is discussed in Section 12.2.
39. To test even moderately complex systems requires an "astronomical" experimental effort (Peters, 1991: 121).
40. Stenseth (1981). To our mutual benefit, Nils Stenseth and I have argued in our homes, at meetings, and in the literature: see Warskowa-Dratnal and Stenseth (1985) for criticism of my ideas, see Chitty (1987) for my criticism of their approach, see Stenseth and Ims (1993) for their rejection of my ideas, and see the present book for my rejection of their rejection. Hairston (1989) notes that for many years "the ecological world stood bemused by mathematical theory as a way of explaining observations made in the field." Perhaps, like Lord Kelvin, theorists consider "that data

that cannot be quantified are hardly worthy of a scientist's attention" (Hallam, 1983: 172). First, however, one must discover which variables are relevant.

41. According to Levin (1981), though models can sometimes be used for prediction, they should more often be regarded as guides to thought, explanation, and experimental design; their metaphorical role should not be mistaken for material truth.

42. Anderson (1960: xiv).

43. Kingsland (1985) and *in lit.*

44. *Oikos* 52(2), 1988.

Section 1.5

45. Whewell (1779–1866) as quoted by Medawar (1967: 32). This is one of several quotations not checked to the original, a sloppy habit discouraged by all good authors, supervisors, and editors. My reasons for breaking my own rule are (1) I trust the secondary sources, which (2) are more accessible to students than the originals.

46. Conant in Mackay (1991). Conant here stretches the meaning of 'proved,' but is sound on the need to give young people freedom to choose their own lines of enquiry.

47. Whitehead (1953: 9).

48. Medawar (1964): *Is the Scientific Paper a Fraud?* "This broadcast was followed by a correspondence . . . illustrating the style of thought that makes scientists treat the 'philosophy of science' with 'exasperated contempt' (Medawar, 1967: 151). A statement by I.B. Cohen (in Green, 1957) applies equally to Bernard and Medawar, that perhaps "sound philosophical analysis of science can be made only by a practising scientist and the greater the scientist the better." Medawar would nevertheless, I think, have approved of the way Hull (1988) first immersed himself in the daily life of a number of scientists before throwing a bright philosophical light on their internecine sociobiology.

49. Dubos (1950: 374).

50. In reviewing a long-term study of Lack (1966), I wrote that "something must be wrong if the present 19 years of data have failed to eliminate even one of the two hypotheses with which the study began" (Chitty 1967b). I also awarded Lack (1964) a prize for one-upmanship for his statement that, as the study was (then) only in its 17th year, it was still too early for him to draw firm conclusions. The ecological mills grind slowly.

51. Beveridge (1950: 17). An experimental tradition is stronger elsewhere in ecology than in the fields discussed in the present book (Paine, 1994).

52. Hull (1973: 46).

53. Huyghens (1629–1695) as quoted by Kneale (1949: 98).

54. Hairston (1989).

55. See p. 181.

56. See p. 54.

Section 1.6

57. Wigglesworth (1961) after I have twice replaced 'the recognition of' by the gerund. *Using* gerunds is shorter and neater than *the use of* gerunds. "Authors who suffer from gerund phobia were presumably once frightened by a dangling participle and are now frightened by all words ending in -ing" (Richards, 1967).

58. Harvey (1578–1657); see Franklin (1958: 29).

59. Elton et al. (1931).

60. Holsinger (1981).

61. Chitty (1952). Years later I was relieved to read that "In science, 'weasel words' serve an important positive function. They buy time while the scientists develop their positions" (Hull, 1988: 7).

Section 1.7

62. Voipio (1950a, b); Voipio (1988).

63. Chitty (1960). To claim that all species are capable of limiting their own population density implies that a mechanism has evolved through natural selection that allows individuals to respond to their neighbors in ways that influence their own survival, reproduction, and dispersal. One consequence of this genetic or phenotypic plasticity is to prevent indefinite increase in numbers. The term 'self-regulation' implies that, within a given set of extrinsic variables, the process is intraspecific.

64. Hansson (1987, 1988), Henttonen et al. (1987).

65. Chitty (1952).

66. Chitty (1967a).

67. Chitty (1987).

68. Chitty (1960). See also Ostfeld (1985).

69. See Section 12.1 for a rejection of the idea that natural selection is a necessary condition for cycles, and see Schaffer and Tamarin (1973) and Gaines (1981) for arguments in favor.

70. Smith (1967).

71. Chitty and Phipps (1966).

72. Chitty (1987).

73. Jenkins et al. (1967): surplus birds were expelled in "one to three sudden stages . . . mortality was a consequence of social behaviour."

74. Curves *abc, afg* and intervening changes in the slopes of the lines can be constructed from Cary and Keith (1979, Table 8) and Keith and Windberg (1978, Tables 7 and 8).

75. For the evidence of early effects on reproduction and survival, see Cary and Keith (1979, Table 1) and Keith and Windberg (1978, Tables 8, 23, and 36). Among other species, dispersal provides further evidence of the early effects of spacing behavior, as it occurs well before peak densities are reached; it is known as 'presaturation dispersal' (Lidicker, 1975; Stenseth and Lidicker 1992). See also Beacham (1980) and Myers and Krebs (1971), who found genetic differences between residents and dispersers.

76. Keith and Windberg (1978: Table 20, Figure 8, p. 37, and p. 50) point out that "poorest reproduction followed highest winter to spring weight losses," which they interpret as indicating a 'critical nutritional phase.' According to my interpretation, by contrast, both loss of weight and poor reproduction are effects of common antecedent conditions, namely, hostilily associated with spacing behavior.

77. Whitehead (1942: 188).

78. Lawton (1992).

79. Crick (1988: 141).

80. Sinclair (1989).

Chapter 2. Pioneering Observations, 1929–1939

1. Whitehead (1950: 157).

2. Elton et al. (1931) and Section 2.7.

3. Middleton (1930, 1931).

4. Elton (1942: 176).

5. David H.S. Davis was my predecessor at the Bureau.

6. It is customary in scientific writing to give metric equivalents of weights and measures originally quoted in English units (and v.v.). I find the resulting clutter almost as distracting as if one were to edit Robert Frost in the way Anthony Lane satirized the process in *The New Yorker,* June 27–July 4, 1994:

> And miles (multiples of 1.6 kilometers)
> to go before I sleep . . .

Conversion factors are given in the index.

Section 2.2

7. Darwin (1888) in Beveridge (1950: 84) from *Life and Letters of C. Darwin.* I have been unable to trace this reference, but it's consistent with Darwin's statement that Spencer's deductive manner of treating every subject was wholly opposed to his frame of mind (Barlow 1958: 109).

8. Summerhayes (1941) gave the trace indices for Newcastleton and Corris, whose trace index for 1933 differed from the trap index in falling lower in spring and recovering less in autumn, but both indices had been at a maximum in autumn 1932. (The published figure for spring 1936 is in error by about half its true value.)

9. Elton (1942: 176).

10. Elton et al. (1935), Elton (1942: 195).

11. *Ibid.*

12. In a study now nearing the end of its 25-year span, Lowell Getz (*in lit.*) and his colleagues have discovered a wide variety of patterns of change in numbers of two species of voles in three habitat types. The conclusion from the first 14 years was that "distinct multiannual population cycles were not characteristic of either species in any habitat" (Getz et al., 1987). Explaining the patterns has been more difficult than describing them.

13. Elton (1942: 186). At this stage, the object was to discover cyclic declines suitable for testing Charles's epidemic hypothesis.

Section 2.3

14. Medawar (1957: 73).

15. Baker et al. (1938); Medawar (1986: 58).

16. Baker and Ranson (1933).

17. Elton et al. (1935).

Section 2.4

18. Beveridge (1950: 111).

19. Sheail (1987).

20. Summerhayes (1941).

21. Elton (1955a).

22. Morris (1963).

23. Fischlin and Baltensweiler (1979).

24. Baltensweiler (1993b).

25. Baltensweiler (1993a).

26. Baltensweiler (1968); Baltensweiler et al. (1977).

Section 2.5

27. Harvey (1578–1657). See Franklin (1957:44).
28. Bacon in Anderson (1960: 60).
29. Hurlburt (1984).
30. Elton et al. (1935).
31. Elton (1942: 192).
32. Goddard (1935).
33. Chitty (1938a).
34. Popper (1963: 53).
35. Geyl (1955).
36. Theorists can invent predators that produce cycles in numbers of their prey:

These conclusions, being implicit in the premises, cannot, of course, be falsified; but through no fault of the mathematicians they have been applied to nature without sufficient regard for important structural dissimilarities between the concrete and abstract systems. The very precision of the models seems to have led to the common error of supposing that their material truth might be assumed . . . and unless the role of mathematics is made clear to the biologist it seems likely that more harm than good may be done. The importance of the models is that they are consistent with the belief that certain types of population change are related to biotic processes, and not necessarily to changes in the physical environment. (Chitty 1957)

Krebs (1985) gives references to these theories and states that "no one has yet found a classic predator-prey oscillation in field populations."

Authors can also describe predators that don't produce cycles in numbers of their prey, and by a judicious combination with those that do, can run the gamut of explanations from cyclic to stationary states (Hanski et al., 1991). These authors observe that small mammals in Fennoscandia have cycles that are longer in the north, more regular, more pronounced, and more in step among species than in the south, where "there is no clear multiannual cycle below 60° N." Correlated with this gradient is a decrease in the number of generalist predators from south to north, where specialist predators, mainly weasels, are said to 'drive' the cycle.

This model "demonstrates that generalist predators have a stabilizing effect on a cycle driven by specialist predators" and, naturally enough, 'predicts' the observations from which it was derived. These conclusions are said to be relevant for the dynamics of cyclic populations in general, but have yet to be tested on independent data to see whether they are consistent with males and females disappearing suddenly at different times, differential survival of young and old, and differences between phases in growth rates and body weights. Moreover, there is so far no evidence to show that cycles (unless increase is prevented altogether) are affected fundamentally by the kind and number of predators.

May (1981) has gone one better than Hanski et al. by producing equations that predict chaotic as well as cyclic and noncyclic states of abundance, the pattern depending only on the value of the intrinsic rate of increase.

37. Erlinge et al. (1990).
38. Krebs et al. (1995). It's no surprise to find that populations increase in density when given unlimited food. But as the increase may be as little as two- to threefold (Gilbert and Krebs, 1981; Boutin, 1990), some surprise may be felt by anyone who believes food is *the* limiting factor instead of one among many. The present 11–fold increase in density is exceptional and leads me to think it was partly a 'Krebs fence effect.' If hares can learn to avoid predators by avoiding open spaces (Hik, 1995),

they can probably learn, when given excess food, to avoid predators lurking in the space outside a fence.

39. Erlinge (1987); Erlinge et al. (1990).
40. Southern (1970).
41. Kidd and Lewis (1987).
42. Moss and Watson (1991).

Section 2.6

43. Bacon in Anderson (1960: 51).
44. Elton (1924).
45. Elton (1942: 190).

Section 2.7

46. Darwin (1859: 73).
47. Elton et al. (1931).
48. Findlay and Middleton (1934); Elton et al. (1935).
49. Elton (1942: 200).
50. Pathological work in the first two studies was done by Dr. G.M. Findlay at the Welcome Institute in London.
51. Chitty (1938b).
52. Chitty (1954a).
53. Frenkel and Friedlander (1951).
54. Chitty (1954a).
55. Elton et al. (1931).
56. Further implication of a faulty water supply is as follows. One of my first jobs at the Bureau was to devise a system that would enable a pathologist to find each vole right after it had died. Some months after I had installed a treadle arrangement wired electrically to call boards salvaged from the *Queen Mary*, a former Atlantic liner, Dr. Wells abandoned my system in favor of a simpler one. He found dead voles by looking for untouched water bottles, correctly reasoning that dead voles can't drink. But voles that can't drink are sure to die; so which was cause and which effect is something we'll never know. I suspect the worst.
57. Koch (1884) in Brock (1961: 116).
58. Baltensweiler and Fischlin (1988).
59. Anderson and May (1980).
60. Anderson and May (1978); May and Anderson (1978).
61. Hudson and Dobson (1990).

Section 2.8

62. Darwin in Barlow (1958: 141).
63. *Ibid,* p. 123.
64. Chitty (1952, 1954a).
65. Wells (1937). See also Wells (1946).
66. Wells (1949).
67. Chitty (1954a).
68. Chamberlin (1890).
69. Elton (1942: 201).

Section 2.9

70. Medawar (1957: 76).

71. Popper (1959: 50). Ayer (1960: 94) sides with Popper: "A man can always sustain his convictions in the face of apparently hostile evidence if he is prepared to make the necessary ad hoc assumptions."

72. 'Cause' is a convenient word in many contexts, but I avoid it where possible in favor of 'necessary conditions' or 'sufficient conditions.' Many authors fail to specify whether they mean a 'necessary' or 'sufficient' cause or realize the difference between them. See Section 6.3 for Medawar's discussion of 'cause'. The argument in the present section is modified from Cohen and Nagel (1934).

73. Bernard in Green (1957: 56): ". . . the only proof that one phenomenon. . . [causes] another is by removing the first, to stop the second."

74. Darwin in Barlow (1958: 84).

75. Platt (1964) writes as follows: "how can we learn the method and teach it? It is not difficult . . . this kind of thinking is not a lucky knack but a system that *can* be taught." Like Bacon, Platt seems to assume that inductive elimination "leaves but little to individual excellence." To a limited extent his remarks may apply to the destructive comments anyone can make in rejecting another person's claim to have discovered something. But how to make constructive discoveries is *not* a system that can be taught. Platt is right, however, in insisting on the value of 'strong inference.' (see Section 12.5).

76. Dampier (1948: 465).

77. Darwin in Barlow (1958: 84).

78. De Morgan (1915, Vol. 1: 80).

79. Medawar (1969:54) later admitted that "The act of falsification is not immune to human error."

80. Statistical hypotheses are even trickier to reject, as unlike universal hypotheses they refer to a proportion instead of to all members of a defined class. As with any proposition referring to some members of a class, its predictions cannot, even in principle, be falsified; they are accepted or rejected with calculable degrees of confidence.

81. Bernard in Green (1957: 17).

82. Conant (1947). Dr. Conant *(in lit.)* was unable to remember where he found this view of Whitehead's; nor have I been able to trace it.

83. Conant (1951).

84. Medawar (1957: 77).

Chapter 3. Qualitative Changes, 1937–39.

1. Crick in Judson (1979: 288).

2. Dymond (ed., 1964).

3. MacLulich (1937).

4. Clarke (1936).

5. Leacock (1961).

6. My own philosophy of education is based on Whitehead (1950: 1, 18), where he says: "A merely well-informed man is the most useless bore on God's earth" and "What education has to impart is an intimate sense for the power of ideas, for the beauty of ideas, and for the structure of ideas, together with a particular body of knowledge which has peculiar reference to the life of the being possessing it." This

philosophy is a good deal harder to put into practice than it is to cram a student's mind with the latest 'facts,' few of which will be so regarded 10 years later.

7. For Medawar's view see Section 1.1, note 2.

8. Conant (1947).

9. Chitty (1936).

10. Chitty (1952).

11. T.H. Huxley recommended 60 as the age at which "men of science ought to be strangled, lest age should harden them against the reception of new truths, and make them into clogs upon progress, the worse in proportion to the influence they had deservedly won" (L. Huxley (1903 LL 2: 418)). I was born in 1912.

12. Von Wright (1957); Hempel (1966).

13. Keith (1974) invoked predation to explain the failure of starvation to account for the continued decline of snowshoe hares. Later on, when starvation turned out to be absent in three out of eight populations, Keith et al. (1984) saved the hypothesis by reinterpreting the role of predation and adding a bad-weather component. He denies (*in lit.*) that his original explanation applies to all declines—which is an admissible disclaimer provided one defines *in advance* which test instances are inadmissable. Pitelka (1973) protected his own views by invoking predation, bad weather, and immigration.

14. Begon et al. (1990). Instead of six factors, Haukioja et al. (1983) settled for a different combination of four: changes in plant availability (qualitative and quantitative), predation, and self-regulation.

Section 3.2

15. Beveridge (1950: 132).

16. Chitty (1952).

17. Overwintered animals present in peak years may have been born later than usual in the previous year of increase and so be younger than usual. They may also have high rates of growth (Chitty 1987).

18. Chitty (1969).

19. Darwin in Mackay (1991).

Section 3.3

20. Bernard in Green (1957: 164).

21. Leslie (1946).

22. Leslie and Ranson (1940).

23. Chitty (1952).

24. Krebs (1979: Fig. 3 and 5).

25. No change in age structure or viability would account for the disappearance within 2 weeks of 96% of a resident population of *M. pennsylvanicus*. Nor could avian predators have been responsible, as the ground was still covered with snow (Mihok et al., 1985).

26. Boonstra (1994).

Section 3.4

27. Kuhn (1962).

28. Green and Larson (1938); Green and Evans (1940); Green et al. (1939).

29. Chitty (1959); includes other references to the Minnesota work.

30. Christian (1950).

31. George Eliot understood better than some ecologists that one cannot refute conclusions merely because they are based on faulty premises. From the apparent falsity of certain premises, Heske et al. (1987) conclude that social behavior does not 'drive' vole cycles and thus commit the fallacy of denying the antecedent (Cohen and Nagel 1934). The authors give failing grades to all hypotheses involving social behavior and A + + to their own ideas about dispersal from habitat patches of different quality.

Section 3.5

32. Crick in Judson (1979: 41).

33. Elton (1942: 201, 190, 205, 202; 204).

34. Facts are by definition true. They should be distinguished from observations, which are always questionable, and from interpretations, which are even more so.

35. According to Conant (1947: 36) "a theory is only overthrown by a better theory, never merely by contradictory facts."

36. Elton (1942: 3).

37. Lidicker (1988).

Chapter 4. Wartime Rat and Mouse Control, 1939–1946

1. Whitehead (1953:3).

2. Hardy (1955). In Mackay (1991). Medawar (1967:122) discusses the snobbery of some 'pure' scientists towards their 'applied' colleagues.

3. Ashby (1959).

4. Elton (1954: 19).

5. Baltensweiler (1993a).

6. Medawar (1991). See also Medawar (1990: 220).

7. Chitty (1954b); Southern (1954).

8. Dubos (1950: 360).

9. Lack (1965).

10. Davis (1987).

11. Calhoun (1949).

12. Christian and Davis (1964).

Chapter 5. Replication, 1946–1951

1. Crick (1988: 126).

2. Stevens (1937).

3. Medawar (1979) has some wise words on women and husband-and-wife teams in science. He also warns those "who go to the extreme length of marrying scientists" to realize beforehand "that their spouses are in the grip of a powerful obsession." Otherwise, they will find out the hard way that their share of running the home is disproportionately great.

4. Watson in Judson (1979: 44).

5. One of the most senseless acts of sabotage was the torching in 1992 of the library and offices of the Institute of Terrestrial Ecology, Banchory, Scotland, probably by animal activists who wished to discourage research on a game bird—despite

the fact that the research was fundamental and not being done for the benefit of the shooting fraternity. Several young professionals and students working for degrees lost over a year's work.

6. Wake (1993).
7. Chitty and Chitty (1962a).
8. Chitty (1938b).
9. Chitty and Kempson (1949).
10. Leslie et al. (1953).
11. Boonstra et al. (1992).

Section 5.2

12. Crick in Judson (1979: 114).
13. Chitty and Chitty (1962b). Male body weights (g) in May 1946–1951 were as follows (those for the peak year in bold type; no visits in May of the low year 1950): 30, 29, **35**, 30, —, 30.
14. Krebs (1979, Figure 3) gives population sizes, and Chitty (1987, Table 3) gives the weight distributions for *M. townsendii*. The animals continued to breed throughout the low phase and to provide emigrants to areas from which all residents had been removed. The unsolved puzzle is know why the animals were so widely spaced in 1971–1974.
15. Krebs (1964a).
16. Years later I discovered that the young grew well if their parents were removed (Chitty 1987).
17. Myllymäki (1977).

Section 5.3

18. Whitehead (1911) in Johnson (1947: 54).
19. Helen was able to distinguish young from old animals of the same body weight "because the inter-orbital region of the skull becomes narrower with age, and the anterior extensions of the temporal ridges become more prominent until they fuse to form a sharp crest. In any one field sample it is possible to make a rough division according to skull characters . . . [In males] the younger-looking skulls nearly always belonged to animals whose seminal vesicles were less well developed than their testes, this time-lag being characteristic of growing animals, at least in *Clethrionomys*. . . . In many cases, however, it was impossible to tell whether a male was young or old, or to distinguish between females, whose placental scars do not persist" (Chitty and Chitty 1962b).
20. Elton et al. (1935); Elton (1942: 140).
21. Cohen and Nagel (1934: 210).
22. Saucy (1988).
23. Krebs (1964a); Pitelka (1973).
24. Cary and Keith (1979) "believe that the correspondence between midwinter-to-spring weight change and reproduction implicates winter nutrition as the primary cause of the cyclic variation." This interpretation is based on correlations late in the cycle. As shown in Section 1.7, however, reproductive changes started before there can have been any lack of good quality food. Food was said to be in short supply throughout the winters of 1970–1972 and perhaps early in the winter of 1972–1973. Nevertheless, body weights of the adults were highest just before, during, and just

after a peak in 1971 and were lowest during decline and scarcity in 1973–1975, when the food supply had recovered. Body weights of yearlings were also lowest in 1973–1975 (Cary and Keith (1979); Keith and Windberg (1978); Pease et al. (1979); Chitty (1987)). In reviewing the snowshoe hare data, Keith (1990) continued to overlook the anomaly of body weights being highest when there was least food. We know of no mechanism to explain these changes in body weight (Chitty 1987), so must at present regard them as symptoms rather than essential characteristics of cycles. In *Clethriono-mys gapperi,* for example, differences between phases are in shape rather than size (Mihok and Fuller 1981).

It is unfortunate that a half-baked suggestion I threw out for discussion at a meeting in Helsinki has been quoted as though I meant it for publication (Taitt and Krebs 1985). I do not now suggest that the heavy animals present at the peak are homozygous dominants.

25. Hik (1995), contrary to my statement that predation does not explain low body weights in snowshoe hares, claimed that it does indeed explain them by preventing hares from feeding in the better habitats. Also, given extra food, they were no longer underweight. Alas, more ugly facts.

Section 5.4

26. Butterfield (1967: 41).
27. Leslie and Chitty (1961).
28. Leslie et al. (1953); see also Chitty and Chitty (1962a).
29. Erlinge et al. (1990), see also Henttonen et al. (1987: Figure 4).
30. Krebs (1979).
31. Hamilton (1937).
32. Elton (1942: 110).
33. In the published version of my thesis (Chitty, 1952) I omitted all reference to Hamilton's study. The shocking truth is that I had filed the reference in the wrong place.
34. Elton et al. (1935).
35. Vogel (1992: 48) says of William Harvey: "He combined a thorough understanding of what his predecessors had found out with an impressive detachment from their conceptual schemes. Any practicing scientist can tell you what a tough combination that is to manage." Yes, indeed.

Section 5.5

36. Schrödinger (1954: 16).
37. Park (1935).

Section 5.6

38. Bernard in Green (1957: 174).
39. Age in days since birth, I realized later, is meaningless when used to compare mature and sexually-inhibited animals.
40. Medawar (1957: 55).
41. Mihok and Boonstra (1992); see Figure 12.3.
42. Leslie et al. (1955).

Section 5.7

43. Bernard in Green (1957: 74).
44. Chitty (1952).
45. See Chapters 2.9 and 6.3 on 'cause.'
46. Hospers (1953).
47. Chitty and Chitty (1962a).
48. Findlay and Middleton (1934).
49. Godfrey (1955). For later examples see Figure 5.12 and Hanski et al. (1993).
50. Hamilton (1937).

Section 5.8

51. Whitehead (1953: 6).
52. Chitty and Chitty (1962a, b).
53. Another characteristic was added by Myers and Krebs (1971). They observed extensive emigration during the increase phase in two species of *Microtus* in Indiana, little or none during the peak and decline. Krebs et al. (1976) and Beacham (1980) confirmed their findings on *M. townsendii* in British Columbia. Mihok et al. (1985), however, observed both emigration and an increase in multiple captures during a decline in numbers of *M. pennsylvanicus*. Chitty (1987) also observed sudden movements during a spring decline in *M. townsendii*.
54. Sandell et al. (1991) doubt the value of a 'cyclicity index' as a way of separating cyclic from noncyclic populations; so does Boonstra (1992); so do I. One of the troubles is that there's no way of telling what degree of scarcity to assign to a catch of zero, in which case the whole data set is useless. Sutton, it may be recalled, sometimes had to walk 4 miles to find pockets of lemmings after a crash and sometimes never found any. Moreover, there's no justification for creating an arbitrary division between pronounced and slight fluctuations.
55. Lidicker and Ostfeld (1991) discuss other influences on body weight.
56. Krebs (1964a).
57. Keith and Windberg (1978).
58. Agrell et al. (1992); see also Nelson et al. (1991).
59. In noncyclic populations of *Clethrionomys gapperi*, spring numbers are usually less than expected from numbers in fall (Mihok 1988).

Chapter 6 Behavior, Physiology, and Natural Selection, 1949–1961

1. Bacon in Anderson (1960: 56).
2. Selye (1936, 1946, 1950).
3. Christian (1950).
4. Chitty (1952).
5. Chitty (1977).
6. Lakatos (1970).

Section 6.2

7. Bernard in Green (1957: 156).
8. Godfrey (1953, 1954).
9. I know of only one way of persuading voles to make nests where you can get

at them; they will often make nests or congregate under sheets of corrugated iron (or any solid cover, I presume) and can be caught by hand if the cover is whipped off smartly enough.

10. Godfrey (1955).

11. Lambin and Krebs (1991).

12. Godfrey (1954). During the decline on the Dell in April-May 1951 Gillian palpated 16 pregnant females and found them carrying 67 embryos. Thus, a decline in catch (from 36 to 9) between May and August was not due to lack of breeding. Meanwhile, the catch increased on Rough Common (from 16 to 100).

13. Despite differences in density and habitats among populations of *Clethrionomys gapperi*, Mihok (1988) found that their population dynamics were nearly identical. Even at low peak densities he observed "massive inhibition of sexual maturity in young of the year."

14. Henttonen et al. (1987) were surprised to find that vole populations may decline in spite of breeding. Elton et al. got over their surprise in 1935. The failure of recruitment in spite of reproduction supports the idea that spacing behavior enables voles to regulate their own rate of increase. On the contrary, according to Henttonen et al. (1987): this is evidence against the hypothesis of self-regulation.

Section 6.3

15. Medawar (1969: 20).

16. Born (1968).

17. Clarke (1956).

18. Clarke (1953).

19. Clarke (1955). The snake pit was so called after a movie of that name starring Olivia de Havilland, a pin-up of John's.

20. Krebs et al. (1969).

21. Spears and Clarke (1988).

Section 6.4

22. Bernard in Green (1957: 126).

23. The stimulating effects of mild interactions between neighbors may be an example of the Allee effect, namely, that things go better with densities in between high and low (Allee, 1931).

24. According to Carter and Getz (1985), reproductive activity in the prairie vole is stimulated by contact between *strangers* of the opposite sex.

25. H. Chitty and Austin (1957).

26. Breed (1967).

27. Scott (1955); Chitty et al. (1956).

28. From a study of 3020 employees of the Boeing Aircraft Company, Bigos et al. (1991, 1992) discovered that back injuries due to emotional distress and dissatisfaction with the job were far commoner than injuries due to physical factors. Back injuries are the most expensive health problem in the industrialized world.

Section 6.5

29. Chitty (1955a).

30. It may be relevant that the female producing this abnormally high proportion

of males (62%) was an unstressed dominant, though one cannot rule out mere chance. The factors affecting sex ratio are complex (Clutton-Brock and Iason, 1986).

31. Carter and Getz (1985) discuss the pheromonal mechanisms by which a mother inhibits the maturation of her offspring

32. Animals 2–8 months of age differed as follows: among the controls, seven of eight males had paired testes over 200 mg; among the experimentals none of 12 had paired testes over 200 mg, and 10 had paired testes less than 100 mg.

33. van Wijngaarden (1960).

34. Frank (1954).

35. Crowcroft (1962).

Section 6.6

36. Beveridge (1950: 49).

37. Delbrück in Judson (1979: 297).

38. Watson in Judson (1979: 20).

Chapter 7. Controversies, 1951–1956

1. Judson (1979: 147)

2. Disraeli (1804–1851). See *The Oxford Dictionary of Quotations,* 3rd ed., 1979.

3. Hull (1988: 22, 32, 26, 222, 288, 228, 228, 115). By stifling discussion, *ad hominem* rudeness does more harm than good, in my opinion; but directed at rival ideas, rudeness is part and parcel of one's professional duties. The line between the two is easily crossed, however, and some egos are very fragile.

Section 7.2

4. This accusation was made about the editorial policy of the *Edinburgh Review* in its early days. The quotation, but not its source, was in my files for 1951.

5. Stenseth and Ims (1993) are wrong in saying it was Elton who rejected this article; he would never have done so merely because it was speculative, nor would he have denied a forum to someone merely because he disagreed with him. His policy with such authors was to give them enough rope to hang themselves.

6. This quotation from A.E. Housman was also in my files and also without a reference.

7. Rick Ostfeld (*in lit.*) makes the good point that a young scientist, in reviewing the manuscript of an older person, is more likely to be honest if his identity is concealed.

8. Medawar (1967: 142).

9. Chitty (1952).

10. Avery in Judson (1979: 39).

Section 7.3

11. Bacon in Anderson (1960: 48, 54).

12. Chitty (1977).

13. Crick (1988: 49).

14. Lack (1965).

Marie Gibbs, secretary to the Bureau from 1935 to 1967, noticed that the question should be rephrased. The bird, a cassowary, did not eat a missionary, it merely wished to do so; its statement was hypothetical not declarative. Also, the side order it had in mind was the man's cassock, not his coat. Had she been an ornithologist, Marie would have pointed out that cassowaries are confined to Australia, New Guinea, and neighboring regions. Her attention to detail and cheerful personality were vital elements in the smooth running of the Bureau. She kept her sanity in spite of the oddballs she had to deal with.

15. Chitty (1967b).
16. *Ibid.*
17. Chitty (1977).
18. Thorpe (1974).
19. Lack (1954a).
20. Lack (1966).
21. Lack (1955), Chitty (1955a).
22. Hull (1988: 222).
23. Ibid.
24. Lack (1954a).
25. Lack (1954a, b; 1955). Henttonen et al. (1987) raised the same objection: ". . . very different densities of a species during successive cyclic peaks normally lead to similar, total crashes . . ." From this they conclude that "density dependent intrinsic mechanisms cannot explain the microtine cycles . . ." in their study area. I agree; the intrinsic mechanisms are not density dependent; they depend on behavior.
26. Lack (1954b).
27. Ibid (also 1954a: 214).
28. Darlington in Chitty (1955a).
29. Lack (1954b). I now agree with David in rejecting my maternal stress hypothesis and accepting a role for natural selection, but Boonstra and Boag (1987) disagree with both of us.
30. Brown (1964) had had much the same idea, and Verner took it up again in 1977. Although I'm still attached to the idea, Hamilton (1970) doubts if an animal would "be ready to harm itself in order to harm another more." Such behavior would reduce the mean fitness of the population and would be uncommon because animals might harm their relatives through being unable to distinguish them from unrelated individuals. (Under disruption, voles beat up both relatives and nonrelatives—but perhaps not equally, see Section 12.4). Rothstein (1979) argues that selection is more likely to improve an individual's own performance than diminish that of its competitors, and Parker and Knowlton (1980) show logically that 'spiteful' behavior (Hamilton's term) must play a smaller role than many authors, including Chitty (1967[a]), seem to think. The ideas of Brown, Chitty, and Verner may well be wrong, but so also may be the assumptions of Rothstein, Hamilton, Parker, and Knowlton.
31. Lack (1954a: 223).
32. Judson (1979: 249); see also Kuhn (1962).
33. Hull (1988: 288) discusses the "alacrity with which scientists seem to be able to misunderstand ideas that seem patent to their authors."

Section 7.4

34. Dampier (1948). The use of reason was regarded as dangerous heresy by Abelard's contemporary, St. Bernard, and by St. Anselm, who, like believers in the doctrine of density dependence, claimed that one must believe in order to understand.

35. T.H. Huxley (1898, CE 1:40).
36. Andrewartha and Birch (1954).
37. Elton (1955b).
38. Medawar (1957: 72).
39. Chitty (1957).

Section 7.5

40. Bondi (1962).
41. Elton (1930: 8).
42. Crowcroft (1991).
43. Chitty (1955b). The following events, reported by Elton and Nicholson (1942a), may have been the ones I thought might be coincidences or less well synchronized than the authors supposed: "The wide synchronization of the cycle in different parts of Canada for at least 100 years, its parallel occurrence both west and east of the Rockies, and its independent occurrence in aquatic species such as the muskrat (*Ondatra zibethica*) and the salmon (*Salmo salar*)."
As well, I contrasted the premises underlying ideas of extrinsic and intrinsic regulation. Krebs (1995a) gives the gist of the argument, which involves a paradigm shift (Kuhn, 1962; Krebs, 1995b) that Charles had not cottoned on to.
44. Fager (1968), using natural and 'synthetic' logs, did a brilliant experimental study of a community of 182 organisms present in decaying wood. "I am particularly grateful to Mr Charles Elton, F.R.S.," he wrote, "who impressed on me the necessity and possibility of work on complex communities and by his interest encouraged me to undertake this study." This study was only one of many that Charles, through the Wytham Ecological Survey, did so much to suggest and make easier.
45. Sheail (1987).
46. von Helmholtz in Beveridge (1950: 68).
47. Touché. Knowing from Bernard Shaw that "reading rots the mind," I feared brain damage from burrowing too deeply into the ecological literature, which I found less intellectually stimulating than reading the history and philosophy of science. Also, according to Beveridge (1950: 3), too much information may smother original ideas and be counterproductive if false. This apprehension puts one in danger, however, of missing stimulating papers, as I did those of Voipio (1950a, b).
48. Cohen and Nagel (1934).
49. Stenseth (1981), recognizing the ambiguity, defined 'relevant' and 'important' as synonymous, which they are not.
50. Elton (1942: 204).
51. Errington (1967).
52. Chitty (1955b).
53. Chitty (1955a).
54. Charles had a good point here: it's easy to become a narrow specialist. On the other hand, it's not easy to make one's way in science without highly concentrated research while working for one's doctorate and during the few years immediately thereafter. Sooner or later one faces the danger of being promoted to a desk-bound job.

Chapter 8. Varying the Circumstances, 1952–1959

1. Whitehead (1922) in Johnson (1947: 55).
2. Middleton (1934, 1935a, b).

3. Ford et al. (1938).
4. Middleton (1935a).
5. Moran (1952).
6. Davis (1987).
7. Lack (1954a).
8. Williams (1954).
9. This and other references in Chitty (1954a).
10. Lack (1954c).
11. Chitty (1977).

Section 8.2

12. Hempel (1966: 34).
13. Lance and Lawton, eds. (1990) show later developments.
14. Jenkins (1961a, b).
15. Jenkins, D. Phil. Thesis (1956).
16. *Ibid.*
17. Myers (1990). See also Myers (1993), Myers and Rothman (1995), and Section 9.3.
18. Jenkins and Watson (1970); Jenkins et al. (1967).
19. "In formal logic, a contradiction is the signal of a defeat: but in the evolution of real knowledge it marks the first step in progress towards a victory." Whitehead (1953: 231).

Section 8.3

20. Medawar (1984a: 24).
21. H. Chitty (1961).
22. H. Chitty and Clarke (1963).

Section 8.4

23. Schrödinger (1954: 58).
24. Chitty (1958).

Section 8.5

25. Bernard in Green (1960: 170).
26. De Morgan (1915, vol. 1:87).
27. Chitty (1958). My remarks on shock disease, offered as an appendix to my talk, were turned down for publication.
28. I abandoned the maternal stress hypothesis in the same year that Christian and Lemunyan (1957) took it up. I still find their evidence unconvincing.
29. Chitty (1958).
30. Whewell in Medawar (1967: 153).
31. Elton (1927: 187).
32. Franz (1953).
33. Watson and Crick (1953).
34. Dawson (1956), Newson (1962).
35. Whitehead (1953: 177).
36. Pareto (1848–1923) in Mackay (1991).

37. Voipio (1950a, b). Serendipidy and logic are equally acceptable *origins* for a hypothesis; neither is adequate for testing it.

38. With acknowledgment to Whitehead's maxim (1926): "Seek simplicity and distrust it."

Section 8.6

39. Hawking (1988: 151).

40. T.H. Huxley in L. Huxley (1903: LL 3:16).

41. Ranson (1934).

42. Leslie et al. (1955).

43. Chitty (1958).

44. Chitty and Phipps (1960).

45. Baker et al. (1963).

46. Medawar (1984b: 31), referring to an early mistake of his own, remarks that "*it doesn't do to be too clever.*"

47. Dubos (1950).

48. Medawar (1969: 1–2).

49. Bernard in Green (1957); also Cohen and Nagel (1934); Section 3.4.

50. "As fallacious as affirming the consequent may be in deductive logic, it is central to science" (Hull 1988: 301).

51. Hull (1973) shows how Darwin at first tried to pass himself off as a good Baconian scientist and how his critics condemned him for not being Baconian enough. According to Medawar (1969: 10–12): the methodology of science "when propounded by scientists is a misrepresentation of what they do" — except presumably of his own methodology.

52. von Helmholtz in Beveridge (1950:60).

53. Whitehead (1953: 9).

Section 8.7

54. Crick (1988: 74).

55. Halle and Lehman (1992) found that voles in Sweden were active by day during a peak, were much less active late in a decline, and switched back to daytime activity during low and early increase phases. My voles that were inactive during the day, were perhaps too early in their increase phase. Nygren (1978) found that voles from the north of Sweden were more active than those from the south, where cycles are said to be absent (but see Figures 2.10, 5.7, 5.12).

56. Chitty (1959).

57. Whitehead (1953: 234): "We do not go about saying that there is another defeat for science because its old ideas have been abandoned."

Section 8.8

58. Bacon in Anderson (1960: 155).

59. Newson and Chitty (1962).

60. In Figure 8.5 male body weights for 1952–1956 are from Chitty and Chitty (1962b), those for 1957–1958 are from Newson & Chitty (1962).

61. Mihok and Boonstra (1992); see Figure 12.3.

62. Evans (1942).

63. Boonstra and Singleton (1993) renewed the search for a physiological explanation for the snowshoe hare cycle. They conclude that unfavorable conditions during the decline may reduce hare reproduction and offspring fitness and thus prolong the low phase.

64. Krebs (1966).

Section 8.9

65. T.H. Huxley (1898, CE 1:62).

66. Chitty (1960).

67. By 1991 the paper had been cited "in more than 205 publications, making it the most-cited paper from [the *Canadian Journal of Zoology*]." The elaboration of these views (Chitty 1967a) also became a Citation Classic, having been cited "in more than 245 publications, making it the most-cited paper from the [*Proceedings of the Ecological Society of Australia*]." (Chitty 1991a, b).

68. Tinbergen (1957). Territorial behavior is only one of the manifestations of hostility and is not a necessary consequence of it.

69. Buxton (1955).

70. Wynne-Edwards (1959, 1962, 1991).

71. Elton (1962).

72. Lack (1966: 311).

73. Maynard Smith (1964); Wiens (1966); Williams (1966); Krebs and Davies (1993).

Chapter 9. From Wytham Woods to Baker Lake, 1960–1962

1. Russell (1931: 34).

2. Bacon in Anderson (1960: 271).

3. Chitty and Phipps (1961, 1966).

4. Andrewartha and Birch (1954). Also Andrewartha (1970), who gives a balanced account of my argument but, naturally enough, rejects my interpretation in favor of his own.

5. Newson (1963).

6. In spite of previous failures to predict changes in population density, I had still not learned that (1) without an estalished theory, prediction is mere 'groping and empiricism,' (2) even with a good theory, prediction is likely to fail in the presence of many uncontrolled variables, and therefore (3) the best one can do is predict *differences in trends* between control and experimental populations. As a way of testing population theory, predicting density itself "is about as decisive as using weather forecasts for testing the laws of physics" (Chitty 1967b).

7. In the middle of May 1960 Wytham males, born in the peak population of 1959, weighed about 26 g; Lake Vyrnwy males, born in an increasing population weighed about 32 g, or about the same as those born during increase in 1958 in Wytham.

Section 9.2

8. Dampier (1948: 133). I have been unable to trace this quotation; but in *The Assayer* (Drake, 1957), Galileo states that the reply "I do not know" is "more beautiful than deceitful duplicity." He also taught his pupils to make a habit of saying "I

do not know"—advice I used to give my own students as a better strategy than bluffing during an oral exam. Schrödinger (1954) repeats the message: "Instead of filling a gap by guesswork, genuine science prefers to put up with it; and this, not so much from conscientious scruples about telling lies, as from the consideration that, however irksome the gap may be, its obliteration by a fake removes the urge to seek after a tenable answer."

9. Pimentel (1958).

10. Birch (1960).

11. Chitty et al. (1968).

12. The fence consisted of sheet metal, 30 in high, enclosing a circular area of 10.85 sq ft. Newsome (1967, 1969) used a similar method with individual house mice in the Australian wheatfields. As food was more concentrated, he enclosed an area of only 1.2 sq m. Moen (1990) did things on a grander scale. His enclosures were 64 sq m, stocked at simulated densities up to 750 voles per ha or 450 lemmings per ha (304 or 182 per acre). Intense grazing during the growing season had little effect on the plant communities. The paper includes references to similar work.

13. To my dismay, when food was provided during the decline of a population of snowshoe hares it prevented loss of body weight (Hik 1995). I'm not sure how to wriggle out of this one—even though the decline continued anyway.

14. In March 1961 males on the feeding area were 3 g lighter than those in the control population and 7 g lighter in April.

15. Populations declined in the presence of excess food given to voles by Krebs and DeLong (1965), by Henttonen et al. (1987) and Erlinge et al. (1990), and given to snowshoe hares by Krebs et al. (1985, 1986a, b). Whether one considers these feeding experiments decisive depends on one's preconceptions. Mine lead me to accept the 1965 results; but Batzli & Pitelka (1971) held firmly to their faith that "nutritive factors could be involved in microtine population cycles." They argued that "neither the adequacy of the natural diet and of supplements in relation to nutrient requirements of *Microtus* nor the effects of feeding stations on social structure were known."

Section 9.3

16. Hawking (1988: 9).

17. Crowcroft (1991).

18. The predictions I made in 1960 were as follows:

(1) If animals are prevented from interacting adversely they should go on increasing until they run out of food. (2) [a] If large enough numbers of animals are continually removed from an expanding population, the survival rate of the remainder should continue to be high; but [b] where an adverse physiological change has occurred no reduction in density should be sufficient to reverse a downward trend. (3) [a] Numbers should continue to increase if animals from an increasing population are sucessfully transferred to an area from which a declining population has been removed; but [b] numbers should continue to decline if animals from a declining population are transferred to a new area.

When I learned that behavior played a direct part in the decline, I got cold feet about the last of these predictions and suggested that a decline due to poor physiological condition might be counterbalanced by behavioral changes on the part of animals introduced into an empty habitat (Chitty, 1987). For the same reason, I now think that 2[b] may be falsified if enough animals are removed from a declining population. In 1960 I thought physiological condition the only relevant variable.

In future experiments cropping should start early in the increase phase. 'Pre-saturation dispersal' (Lidicker, 1975; Stenseth and Lidicker, 1992) suggests that changes in quality may start well before the peak. See also Charnov and Finerty (1980).

19. Smyth (1968).
20. As quoted by Crowcroft (1991).
21. Smyth (1968).
22. Krebs (1966).
23. Krebs (1992b).
24. Hentonnen et al. (1987), Heske et al. (1993).
25. Both 3a and 3b were tested by Krebs & DeLong (1965). See note 15 (above) for references to this and later food experiments.
26. See Section 2.4.
27. Myers (1993) introduced egg masses from different population phases of *Malacosoma pluviale* to unoccupied sites.
28. Krebs (1978).
29. Watson and Moss (1970).
30. Moss and Watson (1990).

Section 9.4

31. In 1865 Bernard wrote that ". . . the greatest confusion reigns, precisely because physicians recognize a multitude of causes for the same disease" (Green, 1957: 83). There are many parallels between population ecology and physiology as it was before the advent of experimental methods. As late as 1925, according to Whitehead (1953: 21), physiology had only just become "an effective body of knowledge, as distinct from a scrap heap. . . . If science is not to degenerate into a medley of *ad hoc* hypotheses, it must become philosophical and must enter upon a thorough criticism of its own foundations."
32. Krebs (1964a).
33. Krebs (1966).
34. Keith (1990: Fig. 16) found that scarring of adult male snowshoe hares dropped from 60% before a peak to 30% at the peak and once again reached 60% at the bottom of the cycle. Scarring of adult females was also higher before the peak and again after it. Juveniles suffered more before the peak than during the decline.
35. Beacham (1970) obtained similar results for *M. townsendii*. Dispersal rates were highest during the spring decline of the peak year and were almost zero during the cyclic decline a year later.
36. Krebs (1964b).
37. Malcolm & Brooks (1993) claim that the changes are purely phenotypic.
38. Chitty (1987).

Chapter 10. Synchrony, 1924–1961

1. Eddington in Judson (1979: 93), who notes that it is a "paradoxical inversion of the inductive system as preached from Bacon to Russell."
2. Chitty (1952). Ydenberg (1987) has since suggested that nomadic predators may affect the timing and onset of a cycle and so bring out-of-phase populations back into phase. To test this idea, Dr. Kai Norrdahl and Dr. Erkki Korpimäki, at the University of Turku, created areas in which avian predation was much reduced. Vole populations, however, continued to fluctuate in synchrony with those in control areas

but with lesser amplitude. This is the only experimental study I'm aware of that tackles the problem of synchrony (Dr. Korpimäki, personal communication).

3. Hik (1995).
4. Elton (1924).
5. MacLulich (1937).
6. Keith (1963).
7. Crowcroft (1991).
8. Elton and Nicholson (1942b).
9. Elton and Nicholson (1942a).
10. Williams (1954), q.v. for scientific names.
11. *Ibid.*
12. Potts et al. (1984).
13. Williams (1985).
14. Moss and Watson (1991).
15. Hanski et al. (1991).
16. Moran (1953a,b).
17. Arditi (1979).

Section 10.2

18. Rutherford's epigraph is quoted slightly differently by different authors.
19. H. Chitty(1950a)
20. The data for Figure 10.2 are from the Snowshoe Rabbit Enquiry, 1931–1948, which was started at the beginning of recovery from a general decline that began in 1924. Notes 1–4 are from MacLulich (1937); notes 5–6 from Professor Rowan, University of Alberta (in the enquiry for (1933–1934); note 7 from Philip (1939); and note 8 from Otto M. Geist (see text). Note 9 is from the enquiry for 1936–1937 and note 10 from the enquiry for 1946–1948. 1. The last year of great abundance and start of the decline was 1924. 2. Peak years progressed from 1932 in the southeast to 1935 in the north. 3. Numbers were at a peak in 1931 in eastern Canada, but, 4, had not begun to increase in the Yukon and Alaska. 5. The peak in 1924–1925 was much higher than that in 1933–1934, when, 6, hares had increased north of Edmonton and decreased in the Athabasca district. Grouse had disappeared almost completely from the north, often before the hares. 7. The die-off was spread over the years 1926–1932. 8. Hares were still plentiful on Anticosti Island but scarce on the north shore of the Gulf of St. Lawrence; they were still plentiful in most parts of Newfoundland. 9. Dawson City: peak followed by die-off; many corpses seen. Mayo district: more abundant (scarce in 1930 or 1931); Whitehorse: abundant; Buffalo Park: scarce. 10. Decrease was gradual and not synchronous thoughout the Yukon. In Alaska hares had begun to decrease near the Yukon border but not further west.

21. According to Philip (1939) the "die-off" does not necessarily occur simultaneously over the whole territory; subsequent rates of increase vary; and in any given area some peaks of abundance are higher than others. "In 1937, it seemed apparent in most areas in Canada that hare populations were approaching, or had reached, a peak. The considerable lag in the decline of the northwestern populations in relation to those farther eastward, was especially true of those in Alaska, since the decline had not yet affected the Seward and Fairbanks areas."

22. Two hunters shot 28 hares in 2 hours in fall 1939, and in 1940–1941 shot 114 in 7 hours; another shot 16 in 10 minutes. (From replies to the questionnaire.)

23. Keith (1963: 62).

24. As cycles of small mammals and game birds differ in length geographically, the same is likely to be true for snowshoe hares.
25. Sinclair et al. (1993).
26. Smith (1983).
27. Whitehead (1953: 126).

Section 10.3

28. Medawar (1969: 59).
29. Hansson and Henttonen (1988).
30. Chitty (1952).
31. Leslie (1959).
32. Moran (1953b).
33. Medawar (1969: 59).
34. Heske and Bondrup-Nielsen (1990); but see Page and Bergerud (1984).
35. "From the intrinsic evidence of his creation, the Great Architect of the Universe now begins to appear as a pure mathematician." (Sir James Jeans, 1930, *The Mysterious Universe*).
36. Hallam (1983).

Chapter 11. Review, 1923–1961

1. Whitehead (1953: 42). Von Helmholtz likened the way he made discoveries to that of someone obliged to retrace his steps after taking many wrong tracks to the top of a mountain, only to see an easy road he might have taken if he had had the wits to find it (Beveridge 1950: 60).
2. Whitehead (1950: 58).
3. Livingstone (1952). He also stated that historians, economists, sociologists, and archaeologists would be 'justly annoyed' if you called them unscientific. According to Conant (1947), however, "To say that all impartial and accurate analyses of facts are examples of the scientific method is to add confusion beyond measure to the problems of understanding science."
4. "All scientists know of colleagues whose minds are so well equipped with the means of refutation that no new idea has the temerity to seek admittance. Their contribution to science is accordingly very small." Medawar (1957: 74; see also 1967: 142).
5. Coleridge. See *The Oxford Dictionary of Quotations*, 3rd ed. 1979.
6. Chitty (1967a).
7. Chitty (1977).
8. T.H. Huxley in L. Huxley (1903, LL 2: 302).
9. The reader should remember that this is an account of how I myself was persuaded by the evidence; other workers have been more resistant, and Boonstra (1994) has resurrected George's idea.
10. Varley and Gradwell (1960); Varley et al. (1973). As early as 1941, Schwerdtfeger, as quoted by Varley (1949), concluded that "The only common factor is the variability of the factors which cause the population changes." This sectarian point of view did not conform to the view of the world championed by George Varley, who commented that ". . . Schwerdtfeger seems to lapse into obscurantism." I disagreed (Chitty, 1955a).
11. Roland (1988, 1990, 1994).

12. Chitty (1967b).

13. "Next to being right in this world, the best of all things is to be clearly and definitely wrong, because you will come out somewhere." T.H. Huxley (1902 CE 3:174).

14. Zinsser in Beveridge (1950: 49).

15. Mihok (1981).

16. Lambin and Krebs (1991).

17. Moss and Watson (1991); Watson et al. (1994).

18. "In the House mouse the unit of population structure is a closed breeding group of small size . . ." Anderson (1970).

19. Chitty (1987).

20. Boonstra (1977). See Section 9.4 for the evidence that it's dangerous for young lemmings to become sexually mature.

21. Increase in numbers, though assumed to be a necessary condition for subsequent events, is not assumed to affect them in a density-dependent manner. Indeed, "population density is assumed to be only one of many environmental variables affecting population trends and not necessarily the best for predicting them" (Chitty 1967a: 57; 1991b). We need some measure of 'crowding' that takes account of behavior, physiological state, and population density, and recognizes that density in one type of environment is not comparable to that in another.

22. See note 9 above. Once again, an idea I had discarded is enjoying a comeback (Boonstra, 1994).

23. Frank (1954).

24. Voipio (1950a) had suggested that "certain individuals, when the stock collapses, survive better than others." See also Voipio (1988) for a review of ideas, starting with Elton (1924), about changes in the direction of selection among fluctuating populations.

25. Krebs (1992b).

26. Chitty and Phipps (1961).

27. MacArthur and Wilson (1967).

28. Rossiter (1994: 761); see also 1992. Wellington (1957) had already recognized that fluctuations in insect populations are associated with profound changes in quality.

29. Mather (1961).

Section 11.2

30. Popper (1963: 36).

31. Hospers (1953: 17).

32. Medawar (1957: 73).

33. Kitaigorodskii (1975). See Mackay (1991).

34. Popper (1959: 41). See Magee (1973) for a condensation of Popper's views. Statements that are certainly true, such as tautologies, contain little useful information. Useful statements are those that *a priori* seem unlikely to be true because they cover a lot of ground and turn out well in spite of the risks they run. (See also Medawar (1990: 91).

35. The Idols of the Market Place are formed "by the intercourse and association of men with each other," and which ". . . lead men away into numberless empty controversies and idle fancies." Bacon in Anderson (1960: 49).

36. Beveridge (1950: 47).

37. Russell (1931).

38. *Ibid.*

39. Whitehead (1953: 32): "If anything out of relationship, then complete ignorance as to it." This quotation may be the sort of leaden style that so prejudiced
Medawar (1984b: xi) against Whitehead.

40. Okansen and Okansen (1992) also recognize the need to apply Koch's postulates but weaken their argument by concluding that "the general fluctuation syndrome
in lowland populations of voles is totally different from that of highland lemmings."
By picking on differences instead of similarities, the authors reduce the chances of
applying the eliminative methods they so rightly advocate.

41. Conant (1947: 43).

42. Those who take no risks will have their knowledge die with them owing to
"excessive fear of ever making a mistake." (Darwin in Barlow, 1958: 103), "Scientists
know that the struggle for acceptance is likely to be arduous, but they are not prepared for how long the process takes, nor for the amount of misunderstanding and
misconstrual that inevitably ensues . . . Anyone promoting a nonstandard view must
be prepared to write the same paper, fight the good fight, answer the same criticisms
over and over again." (Hull 1988: 288). Hence this book and its repetitions.

43. Bernard in Green (1957: 176).

44. Hull (1988: 280).

45. Whitehead (1953: 23).

46. Lakatos (1970: 158).

47. Gaines et al. (1978).

48. Russell (1931).

49. Medawar (1957: 77).

50. Calvin Sandborn in *The Vancouver Sun,* February 21, 1992.

51. Geyl (1955).

52. Dubos (1950); see Section 1.5.

53. Medawar (1967:87).

54. Watson and Crick (1953).

55. *Peter Larkin, a Rhodes Scholar, joined us in 1946 and worked on moles under
Charles's supervision; but I, too, helped him adjust from working in the Saskatchewan
bush to working in the grounds of His Grace the Duke of Marlborough, where the trees
are planted in the formation of the armies at the battle of Blenheim, 1704. They provided
useful markers for trap lines. Peter looked after my voles when I was away, helped make
the first batch of Longworth traps, and worked with me at Lake Vyrnwy en route to
Liverpool and a job at the University of British Columbia. When our bridge games finished after midnight, Peter used to enter Exeter College through a coal chute in Broad
Street. I am grateful to him for his part in bringing me to UBC.*

56. Brend (1984); Hall (1969). Monsignor Ronald Knox proclaimed Sherlock
Holmes an Oxford graduate. Dorothy L. Sayers claimed him for Cambridge.

57. Whitehead (1950: 157): epigraph in Section 2.1.

58. Sherlock Holmes acted much like a scientist, but was saddled with views about
scientific method that reflected the ambivalent views of the late 19th century. What
Holmes says he did and what he actually did are two different things, his accounts of
his methods being "downright false as scientific theory" (Accardo 1987). See also
Gardner (1981) on "The irrelevance of Conan Doyle."

The quotations that follow are from the *The Complete Sherlock Holmes by Sir
Arthur Conan Doyle:* With *a Preface by Christopher Morley.* Garden City, New
York: Doubleday and Company.

Sherlock Holmes originally claimed that to someone trained to observation and analysis "his conclusions were as infallible as so many propositions of Euclid," that "detection is, or ought to be, an exact science and should be treated in the same cold and unemotional manner," and that guessing "is a shocking habit—destructive to the logical faculty." Scientists, it was thought, actually proved things—a legacy of the success of Newtonian physics. Contrary to Holmes's claims, however, his success was largely due to his powerful imagination; his supposedly objective methods did not work in less gifted hands, as Conan Doyle later made clear in a spoof of Holmes's methods when applied by Dr. Watson (Doyle, 1980).

The professional detectives created by Conan Doyle differed from Sherlock Holmes in their unfailing genius for acting on irrelevant clues. As Holmes explained to Watson: one must "be able to recognize out of a number of facts, which are incidental and which vital." The difficulty often lay in "there being too much evidence. What was vital was overlaid and hidden by what was irrelevant." "Nothing is more deceptive," he claimed than an obvious fact," a maxim reminiscent of the seductive effect of Selye's ideas.

By living in a universe created by Conan Doyle, Holmes had an easier life than that of a scientist, who can never assume that his list of possible explanations is complete. Unlike Darwin, Holmes never doubted the truth of the Baconian principle of exclusion, and believed that "when you have eliminated the impossible, whatever remains, *however improbable,* must be the truth."

Because scientists deal with classes of events rather than with events for their own sake, they have an advantage over detectives in having unlimited opportunities to justify their solutions. That is, if they live so long. But if Holmes failed to solve a mystery, he got no second chance. Thus, he never knew what happened to Mr. James Phillimore, who, "stepping back into his own house to get his umbrella, was never more seen in this world." Nor did he discover what unhinged a well-known journalist and duellist, "who was found stark staring mad with a match box in front of him which contained a remarkable worm said to be unknown to science."

Battles between scientists can be as obsessive as those between Sherlock Holmes and Professor Moriarty but without ending in struggles to the death; nor with encounters like that in a Dedini cartoon in which a gun-toting scientist confronts his rival with the words: "Hoffmeister, you've disproved my theories for the last time!" Fortunately, a good scientist would be able to talk his way out of such trouble by asking his rival to define his terms and by convincing him that disproof is impossible in science.

Holmes mistakenly believed in the Baconian principle of the *tabula rasa* when he said "We approached the case . . . with an absolutely blank mind. . . . We had formed no theories. We were simply there to observe and to draw inferences from our observations." Darwin once said much the same thing but was more accurate when he said he could not resist forming a hypothesis on every subject.

The story of the missing race horse, Silver Blaze, provides the best evidence that what Holmes did was the opposite of what he originally said he did. Except that the events could not be replicated, this story bears a close resemblance to a scientific investigation, though unaccompanied, one hopes, by similar unethical conduct. (In real life, Holmes might have landed in jail for using inside information to bet on the supposedly missing favorite (Doyle, 1984). Unlike Inspector Gregory, Holmes soon cut through the "plethora of surmise, conjecture, and hypothesis" and recognized the significance of observations that others had made but failed to interpret—for example, "the curious incident of the dog in the night-time" ('The dog did nothing in the night-time'. 'That was the curious incident.') "See the value of imagination," said Holmes, demolishing at one blow his claim that detection was an exact, cold, and unemotional science, " . . . We imagined what might have happened, acted upon the supposition and find ourselves justified." Having solved the problem of the missing horse and the death of its trainer, he further justified his solution by making the risky but successful predictions that the trainer had a mistress and had maimed a number of sheep.

Oddly enough, Holmes offers the true, synthetic method of science (Ayer, 1960) as a poor substitute for his own so-called deductive methods, saying: "In the absence of data we must abandon the analytic or scientific method of investigation, and must approach it in the synthetic fashion. In a word, instead of taking known events and deducing from them what has occurred, we must build up a fanciful explanation if it will only be consistent with known events. We can then test this explanation by any fresh facts which may arise" (Doyle, 1980). "If the fresh facts which come to our knowledge all fit themselves into the scheme, then our hypothesis may gradually become a solution." This hope, or delusion, is what keeps research workers going. I myself could not have done without it. "A problem without a solution may interest the student, but can hardly fail to annoy the casual reader," wrote Dr. Watson. Anyone who has read this far will presumably have risked this form of annoyance, which scientists have to put up with all the time. Luckily, there are compensations, because "those who do not know the torment of the unknown cannot have the joy of discovery" (Claude Bernard in Beveridge (1950: 75). 'Torment' is a Gallic exaggeration for 'challenge,' which explains why some of us find it hard to quit. The nonscientist who is also an admirer of Sherlock Holmes will understand this affliction.

Chapter 12. Epilogue, 1961–1995

1. Feyerabend (1975) in Mackay (1991).
2. Hull (1988: 22).
3. Zinsser in Beveridge (1950: 107)
4. Voipio (1950a, b; 1988).
5. The first studies were by Semeonoff and Robertson (1968), Tamarin and Krebs (1969, 1973); Gaines and Krebs (1971), and LeDuc and Krebs (1975).
6. Spears and Clarke (1988).
7. Desjardins et al. (1986).
8. Clarke (1977); Spears and Clarke (1987).
9. Moss et al. (1985) suggested that the oberved selection for social dominance "would be facilitated if pairs mated assortatively." Morris (1971) and Baltensweiler (1993a) observed assortative mating among insect populations. Animals that breed under short daylengths are attracted by their odors to other animals of the same kind (Gorman et al., 1993). This attraction would also tend to encourage assortative mating.
10. Nelson (1987).
11. Vogel (1992).
12. Singleton and Hay (1992) found that maternal effects complicated the genetic analysis of aggression, but that there was much additive genetic variation.
13. Boonstra and Boag (1987).
14. Mihok and Boonstra (1992).
15. Boonstra and Hochachka (manuscript).
16. Those who never believed in the relevance of natural selection accept these conclusions joyfully, whereas I find plenty of reasons against them. The shoe is now on the other foot than it was in Section 9.2, where it was I who accepted the conclusions from an experiment in my favor and opponents who found it unconvincing.

Without partisanship like this, bad ideas would slip by too easily and good ideas would have no guardian angel.

17. Whitehead (1953: 230).

18. Eddington (1935: 81).

19. Among larch budmoths, according to Baltensweiler (1993a), the dark morphs at the start of recovery are homozygous recessives, emerge earlier than the lighter-colored morphs, mate assortatively, and thus form a rapidly increasing proportion of the population. They are selected against during the 4-5-year decline.

20. Mitter and Schneider (1987) conclude that "although we do not regard any conclusion about genetic effects on these cycles to be well founded at present, the budmoth remains a candidate for a genetic influence on outbreak dynamics."

Section 12.2

21. Medawar (1979: 72).

22. As I was finishing my own story, Stenseth and Ims (1993) complicated my task by launching *The Biology of Lemmings,* a 1.5-kg, 683-page, $85 book for which, as editors, they had obtained the experts' latest contributions. This is an excellent book, treats with fairness the rival schools of thought, places my own autobiographical account in its wider context, and cites theoretical reasons against my ideas. Except for referring to a limited number of its contributors, I'm happy to leave future students to sort out differences between its accounts and mine. They may be pleased to learn that, as with Ontario, the truth is still theirs to discover.

23. It may rightly be objeted that having me comment on these studies is equivalent to having General Sherman become fire marshall of Atlanta. (I could not resist adapting this comment from a remark of President Clinton's.)

24. Laine and Henttonen (1983), whose rejection of hypotheses of self-regulation was followed by Jarvinen's rejection of their hypothesis of cyclic plant production. The authors, he suggests, have mistaken effects for causes (Jarvinen (1987).

25. Hilborn and Stearns (1982). For an opposing view see Tamarin (1978).

26. Hansson and Henttonen (1988).

27. Lidicker (1988, 1991), but see Gaines et al. (1991).

28. Section 3.1.

29. Hanski et al. (1991).

Section 12.3

30. Popper (1963: 36).

31. Black quoted by Popper (1959: 82).

32. Pitelka (1958: 250).

33. Pitelka (1973: 199).

34. Pitelka (1964: 55).

35. Schultz (1964: 57).

36. Pitelka (1973: 209).

37. According to Pitelka (1964), the following events explain why expectations from the nutrient recovery hypothesis were not observed. Numbers refer to those in Figure 12.5. 1. At the abundance observed in June, the vegetation would not have been damaged enough to cause a decline. Numbers had probably been higher earlier on—before losses to avian predators. 2. This decline was delayed until late summer and the following winter. Predation by jaegers had been unusually light. 3. The ex-

pected prepeak increase was wiped out by an early July snowstorm, high rainfall, and subnormal temperatures. 4. Lemmings were inaccessible to avian predators because of a persistent regional storm between 5 and 19 June; and as jaegers were scarce, lemmings were more than normally abundant at the end of summer. 5. Lemming numbers were higher than usual after a peak. No weasels were seen. 6. A peculiar freezeup prevented lemmings from foraging normally. This and exceptional numbers of weasels prevented lemmings from reaching the expected peak. 7. An ice glaze in October was followed by a severe winter with little snow cover and below-normal temperatures in January–May. This seond harsh winter plus overcropping by weasels finally ended the decline. 8. Winter nests and young lemmings were scarce, adult females had not been breeding, and the vegetation was in good condition in summer. As these events were not typical of a peak, the high numbers did not "constitute a cyclic peak in the strict sense" and indicate that most brown lemmings had immigrated, perhaps from 95 km away. 9. Six to seven years had elapsed "since the last strong pulse of dead plant matter was added to top soil by lemming grazing;" but in spite of a favorable food supply, the population had declined by June 1972. This decline was due to poor snow cover and abundance of arctic foxes. 10. Collared lemmings suddenly became abundant, evidently because of the sustained low density of the other species. This increase, unlike that in brown lemmings, was due to local breeding.

38. Pitelka (1973: 202).
39. Schultz (1964: 57).

Section 12.4

40. Semeonoff and Robertson (1968).
41. Tamarin and Krebs (1969); Gaines and Krebs (1971).
42. Krebs et al. (1969).
43. Ostfeld (1994) has reexamined (and found wanting) the claim that fenced-in voles always run out of food.
44. Krebs (1979).
45. LeDuc and Krebs (1975).
46. I've never managed to breed deer mice in small cages; but when I let some loose in one of the animal rooms, they reproduced while pairs in the cages still did not.
47. Chitty (1987).
48. Kikkawa (1964).
49. Hamilton (1970). In the excitement of attacking strangers, residents also attacked residents, but perhaps less persistently.

Section 12.5

50. Whitehead (1950: 147).
51. Medawar (1979).
52. Lidicker (1994) gives a list of seven rather different topics suitable for mammalogists to work on. Cockburn (1988) should also be consulted in a book kindly dedicated to me "For agreeing to disagree over whether we agree." Reviews by Batzli (1992) and Krebs (1993) will also prove useful, and between these authors and Stenseth and Ims, eds., (1993), a student should be able to trace points of view I have dismissed too readily or failed to cover (for I'm less even less *au fait* with the literature

than I used to be—Section 7.5). Many authors have helped bridge the gap between the present and 1961, which is where I once meant to end my story, but like Topsy it's "just growed." See also Desy and Batzli (1989) and Hestbeck (1982).

53. Platt (1964).

54. Tamarin (1977); Hansson (1984).

55. Nygren (1978); Nyholm and Meurling (1979); Gustafsson et al. (1983a, b); Hansson (1984); Agrell et al. (1992).

56. Rasmuson et al. (1977); Nygren (1980).

57. See also Halle and Lehman (1992).

58. Sandell et al. (1991); see also Section 5.8.

59. Mihok et al. (1985) give a real-life example of a population switching from a cyclic to a stationary state.

60. The small size of the peak in 1972 on Westham Island was due to the lateness of recovery in 1971 (not shown in Figure 12.7). Twenty percent of its males weighed 70 g or more in March 1972, whereas only one animal weighed this much in the next 2 years. In the high peak of March 1975 56% of the males weighed 70 g or more. Weight distributions are given by Krebs (1979). Body weights at Corris were high in 1932 (see Figure 5.3), but were not followed by a severe decline (see Figure 2.2). As the crow flies, Corris is only 18 miles across the mountains from Lake Vyrnwy, but we had neither time nor money to study both populations.

61. Baltensweiler (1968).

62. Having a runway across my desk (see Figure 6.5) enabled me to be on hand when boundary disputes erupted.

63. Wolff et al. (1983) and Wolff (1985) staged encounters between resident and visiting deer mice inside an open-bottomed plastic cyclinder placed on natural substrate within the residents' home range.

64. Tamarin et al. (1984).

65. Boonstra and Krebs (1977); Beacham (1980); Krebs (1992b); Ostfeld (1994).

66. Krebs et al. (1995).

67. Boonstra and Craine (1986).

68. Krebs (1988) discusses some of the pros and cons of this approach to knowledge.

69. Whitehead (1922): epigraph to Section 8.1.

70. Oreskes et al. (1994) show that models in the earth sciences suffer from defects similar to those in population ecology.

71. Page and Bergerud (1984).

72. Dampier (1948: 105).

73. Eddington (1935: 81).

74. Reviewed by Gaines (1981).

75. Rose and Gaines (1981).

76. Current ideas are that "Phase characters [in locusts] are transmitted from generation to generation by nongenic inheritance through the cytoplasm of the egg . . . These maternal influences might be passed on for several generations." But natural selection may also play a part. See Krebs (1985).

77. Tamarin and Sheridan (1987) made early attempts to determine kinships.

78. Dibble (1993).

79. Moss et al. (1984).

80. Moss and Watson (1991); Watson et al. (1994)

81. Lambin and Krebs (1993).

82. Lakatos (1970).

References

Numbers in parentheses refer to the section(s) in which the reference is cited

Accardo, P. 1987. *Diagnosis and detection. The medical iconography of Sherlock Holmes*. London: Fairleigh Dickinson University Press: Associated University Presses, Inc. (11.2).

Agrell, J., S. Erlinge, J. Nelson, and M. Sandell. 1992. Body weight and population dynamics: Cyclic demography in a noncyclic population of the field vole (*Microtus agrestis*). *Can. J. Zool.* 70: 494–501. (5.8, 12.5)

Allee, W.C. 1931. *Animal aggregations: A study in general sociology*. Chicago: University of Chicago Press. (6.4)

Anderson, F.H., ed. 1960. *Francis Bacon: The New Organon and related writings*. New York: The Liberal Arts Press, (P, 1.3, 1.4, 2.5, 2.6, 6.1, 7.3, 8.8, 9.1, 11.2)

Anderson, J.W. 1939. Summer cruise to the arctic—1939. *The Beaver, Outfit* 270 (3): 44–45. See also *Outfit* 269 (2): 48–52, 1938. (1.3)

Anderson, R.M. and R.M. May. 1978. Regulation and stability of host-parasite population interactions: I. Regulatory processes. *J. Anim. Ecol.* 47: 219–249. (2.7)

Anderson, R.M., and R.M. May, 1980. Infectious diseases and population cycles of forest insects. *Science* 210: 658–661. (2.7)

Anderson, P.K. 1970. Ecological structure and gene flow in small mammals. *Symp. Zool. Soc. Lond.* 26: 299–325. (11.1).

Andrewartha, H.G. 1970. *Introduction to the study of animal populations*. 2nd ed. London: Methuen. (9.1).

Andrewartha, H.G. and L.C. Birch. 1954. *The distribution and abundance of animals*. Chicago: University of Chicago Press. (7.4, 9.1).

Arditi, R. 1979. Relation of the Canadian lynx cycle to a combination of weather variables: A stepwise multiple regression analysis. *Oecologia* 431: 219–233. (10.1).

Ashby, E. 1959. *Technology and the academics. An essay on universities and the scientific revolution*. London: MacMillan. (4).

Ayer, A.J. 1960. *Language, truth and logic*. London: Victor Gollancz Ltd. (2.9, 11.2).

Bacon, F. 1620. In Anderson (1960).

Baker, J.R., D. Chitty, and E. Phipps, 1963. Blood parasites of wild voles, *Microtus agrestis*, in England. *Parasitology* 53: 297–301. (8.6)

Baker, J.R., and R.M. Ranson. 1933. Factors affecting the breeding of the field mouse (*Microtus agrestis*). Part III.—Locality. *Proc. R. Soc. Lond. B.* 113: 486–495. (2.3).

Baker, J.R., R.M. Ranson, and J. Tynen. 1938. A new chemical contraceptive. *Lancet* Oct. 15: 882–885. (2.3)

Baltensweiler, W. 1968. The cyclic population dynamics of the grey larch tortrix *Zeiraphera griseana* Hübner (= *Semasia diniana* Guenée) (Lepidoptera: Tortricidae). In T.R.E. Southwood (ed.) *Insect abundance. Symp. R. Ent. Soc., Lond.* 4: 88–97. (2.4, 9.3, 12.5).

Baltensweiler, W. 1993a. A contribution to the explanation of the larch budmoth cycle, the polymorphic fitness hypothesis. *Oecologia* 93: 251–255. (2.4, 4, 12.1)

Baltensweiler, W. 1993b. Why the larch budmoth cycle collapsed in the subalpine larch-cembran pine forests in the year 1990 for the first time since 1850. *Oecologia* 94: 62–66. (2.4)

Baltensweiler, W., G. Benz, P. Bovey, and V. Delucchi (1977). Dynamics of larch budmoth populations. *Ann. Rev. Entomol.* 22: 79–100. (2.4, 9.3)

Baltensweiler, W. and A. Fischlin. 1988. The larch budmoth in the Alps. In A. Berryman, ed.: *Dynamics of forest insect populations,* pp. 331–351. London: Plenum. (2.7)

Barlow, N. 1958. *The autobiography of Charles Darwin 1809–1882. With original omissions restored. Edited with appendix and notes by his grand-daughter.* London: Collins. (1.1, 2.2, 2.8, 2.9, 11.2)

Batzli, G.E. 1992. Dynamics of small mammal populations: A review. In D.R. McClough and R.H. Barrett, eds. *Wildlife 2001: Populations.* 831–850. New York: Elsevier Applied Science. (12.5)

Batzli, G.O. and F.A. Pitelka. 1971. Condition and diet of cycling populations of the California vole, *Microtus californicus. J. Mammal.* 52: 141–163. (9.2)

Batzli, G.O, R.G. White, S.F. Maclean, Jr., F.A. Pitelka, and B.D. Collier. 1980. The herbivore-based trophic system. In J. Brown, P.C. Miller, L.L. Tieszen, and F.L. Bunnell, eds. *An arctic ecosystem: The coastal tundra at Barrow, Alaska.* US/IBP Synthesis Series 12, pp. 335–410. Stroudsburg, P: Dowden, Hutchinson and Ross, Inc. (1.4)

Beacham, T. D. 1980. Dispersal during population fluctuations of the vole, *Microtus townsendii. J. Anim. Ecol.* 49: 867–877. (1.7, 5.8, 9.4, 12.5)

Begon, M, Harper, J.L., and Townsend, C.R. 1990. *Ecology: Individuals, populations and communities.* Oxford: Blackwell (3.1)

Bell. R. 1992. *Impure science: Fraud, compromise and political influence in scientific research.* New York: Wiley (P)

Bernard, C. 1865. In Green (1957)

Berry, R.J. (1991). Charles Elton FRS; The book that most . . . *Biologist* 38: 127–129. (Ack).

Beveridge, W.I.B. 1950. *The art of scientific investigation.* London: Heinemann (P, Ack, 1.1, 1.5, 2.2, 2.4, 3.2, 6.6, 7.5, 8.6, 11.1, 11.2, 12.1).

Bigos, S.J., M.C. Battie, D.M. Spengler, L.D. Fisher, W.E. Fordyce, T.H. Hansson, A.L. Nachemson, and M.D. Wortley. 1991. A prospective study of work perceptions and psychosocial factors affecting the report of back injury. *Spine* 16: 1–6. (6.4)

Bigos, S.J., M.C. Battie, D.M. Spengler, L.D. Fisher, W.E. Fordyce, T. Hansson, A.L. Nachemson, and J. Zeh. 1992. A longitudinal, prospective study of industrial back injury reporting. *Clin. Orthopaed.* 279: 21–34. (6.4)

Birch, L.C. 1960. The genetic factor in population ecology. *Am. Nat.* 94: 5–24. (9.2).

Birkeland, K. 1986. Staloer tror at månen er et bål: 25 Sorsamiske Eventyr. J.W. Cappelen Forlag A.S. (1.2)

Bondi, H. 1962. Why scientists talk. *Adv. Sci.* September: 259–267. (1.1, 7.5)

Boonstra, R. 1977. Effect of conspecifics on survival during population declines in *Microtus townsendii. J. Anim. Ecol.* 46: 835–851. (11.1).

Boonstra, R. 1992. Measuring temporal variability of population density: a critique. *Am. Nat.* 140: 883–892. (5.8).

Boonstra, R. 1994. Population cycles in microtines: The senescence hypothesis. *Evol. Ecol.* 8: 196–219. (3.3, 11.1).

Boonstra, R. and P.T. Boag, 1987. A test of the Chitty Hypothesis: Inheritance of life-history traits in meadow voles *Microtus pennsylvanicus. Evolution* 41: 929–947. (7.3, 12.1).

Boonstra, R. and I.T.M. Craine. 1986. Natal nest location and small mammal tracking with a spool and line technique. *Can. J. Zool.* 64: 1034–1036. (12.5)

Boonstra, R. and W.M. Hochachka. (MS). Maternal and additive genetic inheritance, and population dynamics of collared lemmings (Dicrostonyx kilangmiutak). (12.1).

Boonstra, R., M. Kanter, and C.J. Krebs. 1992. A tracking technique to locate small mammals at low densities. *J. Mammal* 73: 683–685. (5.1)

Boonstra, R. and C.J. Krebs, 1977. A fencing experiment on a high-density population of *Microtus townsendii. Can. J. Zool.* 55: 1166–1175. (12.5).

Boonstra, R. and G.R. Singleton. 1993. Population declines in the snowshoe hare and the role of stress. *Gen. Comp. Endocrinol.* 91: 126–143. (8.8).

Born, M. 1968. *My life and views.* New York: Charles Scribner's Sons. (6.3).

Boutin, S. 1990. Food supplementation experiments with terrestrial vertebrates: patterns, problems and the future. *Can. J. Zool.* 68: 203–220 (2.5)

Breed, W.G. 1967. Ovulation in the genus *Microtus. Nature* 214: 826. (6.4).

Brend, G. 1984. Oxford or Cambridge. In Shreffler (1984), q.v., pp. 109–115 (11.2)

Brock, T.D. (ed.) 1961. *Milestones in microbiology.* Englewood Cliffs, N. J.: Prentice-Hall, (2.7).

Brown, J. L. 1964. The evolution of diversity in avian territorial systems. *Wilson Bull.* 76: 160–169. (7.3).

Butterfield, H. 1967. *The origins of modern science 1300–1800.* New York: The Free Press. (5.4)

Buxton, P.A. 1955. The natural history of tsetse flies. An account of the biology of the genus *Glossina* (Diptera). *Mem. Lond. School Hyg. Trop. Med.* 10: 1–816. (8.9)

Calhoun, J.B. 1949. A method for self-control of population growth among mammals living in the wild. *Science* 109: 333–335. (4)

Carter, C.S. and L.L. Getz. 1985. Social and hormonal determinants of reproductive patterns in the prairie vole. In R. Gilles and J. Balthazart, eds. *Neurobiology,* pp. 18–36. Berlin, Springer, (6.4, 6.5)

Cary, J.R. and L.B. Keith. 1979. Reproductive change in the 10-year cycle of snowshoe hares. *Can. J. Zool.* 57: 375–390. (1.7, 5.3)

Chamberlin, T.C. 1890. The method of multiple working hypotheses. Reprinted in *Science* 148: 754–759 (1965). (2.8)

Charnov, E.L. and J.P. Finerty. 1980. Vole population cycles. A case for kin-selection? *Oecologia* 45: 1–2. (9.3)

Chitty, D. 1936. A ringing technique for small mammals. *J. Anim. Ecol.* 6: 36–53. (3.1)

Chitty, D. 1938a. A laboratory study of pellet formation in the short-eared owl *(Asio flammeus). Proc. Zool. Soc. Lond.* A 108:267–287. (2.5)

Chitty, D. 1938b. Live trapping and transport of voles in Great Britain. *J. Mammal.* 19: 65–70. (2.7, 5.1)

Chitty, D. 1952. Mortality among voles (*Microtus agrestis*) at Lake Vyrnwy, Montgomeryshire in 1936–9. *Philos. Trans. R. Soc. Lond.* [Biol.] 236: 505–552. (1.6, 1.7, 2.8, 3.1, 3.2, 3.3, 5.4, 5.7, 6.1, 7.2, 10.1, 10.3)

Chitty, D. 1954a. Tuberculosis among wild voles: With a discussion of other pathological conditions among certain mammals and birds. *Ecology* 35: 227–237. (2.7, 2.8, 8.1)

Chitty, D., ed. 1954b. *Control of rats and mice. Vols. 1 and 2: Rats.* Oxford: Clarendon. (4)

Chitty, D. 1955a. Adverse effects of population density upon the viability of later generations. In J.B. Cragg and N.W. Pirie, eds. *The numbers of man and animals.* pp. 57–67. London: Oliver and Boyd. (6.5, 7.3, 7.5, 11.1)

Chitty, D. 1955b. Allegemeine gedankengänge über die Dichteschwankungen bei der Erdmaus (*Microtus agrestis*). *Z. Säugetierkunde* 20: 55–60. (7.5)

Chitty, D. 1957. Population studies and scientific methodology. *Br. J. Philos. Sci.* 8 (29) 64–66. (2.5, 7.4)

Chitty, D. 1958. Self-regulation of numbers through changes in viability. *Cold Spring Harbor Symp. Quant. Biol.* 22 [1957]: 277–280. (8.4, 8.5, 8.6)

Chitty, D. 1959. A note on shock disease. *Ecology* 40: 728–731. (3.4, 8.7)

Chitty, D. 1960. Population processes in the vole and their relevance to general theory. *Can. J. Zool.* 38: 99–113. (1.7, 8.9, 9.3)

Chitty, D. 1964. Animal numbers and behaviour. In J.R. Dymond, ed. *Fish and wildlife: A memorial to W.J.K.Harkness,* pp. 41–53. Toronto: Longmans, (3.4)

Chitty, D. 1967a. The natural selection of self-regulatory behaviour in animal populations. *Proc. Ecol. Soc. Aust.* 2: 51–78. (1.7, 7.3, 8.9, 11.1)

Chitty, D. 1967b. What regulates bird populations? *Ecology* 48: 698–701. (1.5, 7.3, 9.1, 11.1)

Chitty, D. 1967c. Review of Crowcroft (1966). (Ack)

Chitty, D. 1969. Regulatory effects of a random variable. *Am. Zool.* 9: 400. (3.2)

Chitty, D. 1977. Natural selection and the regulation of density in cyclic and noncyclic populations. In B. Stonehouse and C. Perrins eds. *Evolutionary ecology,* pp. 27–32. London: Macmillan, (6.1, 7.3, 8.1, 11.1)

Chitty, D. 1987. Social and local environments of the vole *Microtus townsendii. Can. J. Zool.* 65: 2555–2566. (1.4, 1.7, 3.2, 5.2, 5.3, 5.8, 9.3, 9.4, 11.1, 12.4)

Chitty, D. 1991a. Vole populations: A model for others? *Citation Classics* 20, May 20: 10. (8.9)

Chitty, D. 1991b. An alternative to density-dependence. *Citation Classics* 22, June 3: 8. (8.9, 11.1)

Chitty, D. and H. Chitty 1962a. Population trends among the voles at Lake Vyrnwy, 1932–60. *Symp. Theriologicum, Brno, 1960:* 67–76. (5.1, 5.4, 5.7, 5.8)

Chitty, D., H. Chitty, P.H. Leslie, and J.C. Scott. 1956. Changes in the relative size of the nucleus in the intervertebral discs of stressed Orkney Voles (*Microtus orcadensis*). *J. Pathol. Bact.* 72: 459–470. (6.4)

Chitty, D. and D.A. Kempson. 1949. Prebaiting small mammals and a new design of live trap. *Ecology* 30: 536–542. (5.1)

Chitty, D. and E. Phipps 1960. The effect of fleas on spleen size in voles. *J. Physiol.* (Lond.) 151: 27–28. (1.7, 8.6)

Chitty, D. and E. Phipps 1961. A declining vole population. *J. Anim. Ecol.* 30: 490–491. (9.1, 11.1)

Chitty, D. and E. Phipps 1966. Seasonal changes in survival in mixed populations of two species of vole. *J. Anim. Ecol.* 35: 313–331. (1.7, 9.1)

Chitty, D., D. Pimentel, and C.J. Krebs. 1968. Food supply of overwintered voles. *J. Anim. Ecol.* 37: 113–120. (9.2)

Chitty, D. and M. Shorten. 1946. Techniques for the study of the Norway rat (*Rattus norvegicus*). *J. Mammal.* 27:63–78. (4)

Chitty, H. 1950a. The Snowshoe Rabbit Enquiry, 1946–48 [with references to previous reports]. *J. Anim. Ecol.* 19:15–20. (10.2)

Chitty, H. 1950b. Canadian Arctic Wild Life Enquiry, 1943–49: with a summary of results since 1933. *J. Anim. Ecol.* 19: 180–193. (1.2, 1.3).

Chitty, H. 1961. Variations in the weight of the adrenal glands of the field vole, *Microtus agrestis*. *J. Endocrinol.* 22: 387–393. (8.3)

Chitty, H. and C.R. Austin 1957. Environmental modification of oestrus in the vole. *Nature* 179: 592–593. (6.4)

Chitty, H. and D. Chitty. 1962b. Body weight in relation to population phase in *Microtus agrestis*. *Symp. Theriologicum, Brno* 1960: 77–86. (5.2, 5.3, 5.8, 8.8)

Chitty, H. and J.R. Clarke 1963. The growth of the adrenal gland of laboratory and field voles, and changes in it during pregnancy. *Can. J. Zool.* 41: 1025–1034. (8.3)

Christian, J.J. 1950. The adreno-pituitary system and population cycles in mammals. *J. Mammal.* 31: 247–259. (3.4, 6.1)

Christian, J.J. and D.E. Davis. 1964. Endocrines, behavior, and population. *Science* 146: 1550–1560. (4)

Christian, J.J. and C.D. Lemunyon. 1957. Adverse effects of crowding on reproduction and lactation of mice and two generations of their progeny. *Naval Med. Res. Inst. Res. Rep.* NM 24 01 00.04.01, 15, 925–936. (8.5).

Clarke, C.H.D. 1936. Fluctuations in ruffed grouse, *Bonasa umbellus* (Linné) with special reference to Ontario. *Univ.Toronto Stud. Biol.* Ser. 41: 1–118. (3.1)

Clarke, J.R. 1953. The effect of fighting on the adrenals, thymus and spleen of the vole *(Microtus agrestis)*. *J. Endocrinol.* 9: 114–126. (6.3)

Clarke, J.R. 1955. Influence of numbers on reproduction and survival in two experimental vole populations. *Proc. R. Soc. Biol. Sci.*, 144: 68–85. (6.3)

Clarke, J.R. 1956. The aggressive behaviour of the vole. *Behaviour* 9: 1–23. (6.3)

Clarke, J.R. 1977. Long and short term changes in gonadal activity of field voles and bank voles. *Oikos* 29: 457–467. (12.1)

Clutton-Brock, T.H. and G.R. Iason. 1986. Sex ratio variation in mammals. *Q. Rev. Biol.* 61: 339–374. (6.5)

Cockburn, A. 1988. *Social behaviour in fluctuating populations*. New York: Croom Helm/Methuen. (12.5)

Cohen, M.R. and E. Nagel. 1934. *An introduction to logic and scientific method.* London: Routledge and Kegan Paul Ltd. (2.9, 3.4, 5.3, 7.5, 8.6)

Coleridge, S.T. 1831. *Oxford Dictionary of Quotations*, 3rd ed., 1979. London: Oxford University Press. (11.2)

Committee of Inquiry on Grouse Disease. 1911. *The grouse in health and in disease: Being the final report of the Committee of Inquiry on Grouse Disease*. London, 2 vols. (8.1)

Conant, J. B. 1945. In Mackay (1991). (1.5)

Conant, J.B. 1947. *On understanding science: an historical approach*. New Haven: Yale Univ. Press, (2.9, 3.1, 3.5, 11.1, 11.2)

Conant, J.B., 1951. *Science and common sense*. London: Oxford University Press. (2.9)

Crick, F. 1988. *What mad pursuit: A personal view of scientific discovery.* New York: Basic Books. (P, Ack, 1.7, 5.1, 7.3, 8.7)

Crick, F.C. 1994. *The astonishing hypothesis: The scientific search for the soul.* New York: Scribner. (P)

Crowcroft, P. 1962. Relating the laboratory environment to nature. *Lab. Anim. Centre Coll. Pap.* 11: 90–16. (6.5)

Crowcroft, P. 1966. *Mice all over.* Chester Springs, PA.: Dufour, (Ack)

Crowcroft, P. 1991. *Elton's ecologists: A history of the Bureau of Animal Population.* Chicago: University of Chicago Press. (Ack, 7.5, 9.3; 10.1)

Dampier, W.C. 1948. *A history of science and its relations with philosophy and religion.* Cambridge: Cambridge University Press. (2.9, 7.4, 9.2, 12.5)

Darlington, C.D. 1955. Discussion. In Chitty (1955a). (7.3)

Darwin, C. 1859. *The origin of species by means of natural selection . . .* 6th ed. London: J.M. Dent and Sons, Ltd., 1928 ed. (2.7)

Davis, D.E. 1987. Early behavioral research on populations. *Am. Zool.* 27: 825–237. (4, 8.1)

Dawson, J. 1956. Splenic hypertrophy in voles. *Nature* 178: 1183–1184. (8.5)

De Morgan. 1915. A budget of paradoxes. (D.E. Smith, ed.) 2nd ed. 2 vols. Chicago: Open Court Publ. Co. (2.9, 8.5)

Desjardins, C., F.H. Bronson, and J.L. Blank. 1986. Genetic selection for reproductive photoresponsiveness in deer mice. *Nature* 322: 172–173. (12.1)

Desy, E.A. and G.O. Batzli. 1989. Effects of food availability and predation on prairie vole demography: A field experiment. *Ecology* 70: 411–421. (12.5)

Dibble, R.O. 1993. Lifetime reproductive success and its correlates in the monogamous rodent, *Peromyscus californicus. J. Anim. Ecol.* 61: 457–468. (12.5)

Doyle, A.C. 1980. *Sherlock Holmes: The published apocrypha. Selected and edited by Jack Tracy.* Boston: Houghton Mifflin (11.2)

Doyle, A.C. 1984. Sidelights on Sherlock Holmes. In P.A. Shreffler, ed. *The Baker Street reader: cornerstone writings about Sherlock Holmes,* pp. 11–16. Westport, CT: Greenwood Press. (11.2)

Drake, S. 1957. *Discoveries and opinions of Galileo.* New York: Doubleday, (9.2)

Dubos, R.J. 1950. *Louis Pasteur: Free lance of science.* Boston: Little, Brown and Company. (1.5, 4, 8.6, 11.2)

Dunbar, M.J. 1980. The blunting of Occam's Razor, or to hell with parsimony. *Can. J. Zool.* 58: 123–128. (1.1)

Dymond, J.R., ed. 1964. *Fish and wildlife: A memorial to W.J.K. Harkness.* Toronto: Longmans Canada Ltd. (3.1)

Eddington, A. 1935. *New pathways in science.* Cambridge: Cambridge University Press (10.1, 12.5)

Eddington, A.S. 1935. The nature of the physical world. London: J.M. Dent and Sons. (12.1)

Elton, C.S. 1924. Fluctuations in the numbers of animals: Their causes and effects. *Br. J. Exp. Biol.* 2: 119–163. (2.6, 10.1, 11.1)

Elton, C. 1927. *Animal ecology.* London: Sidgwick and Jackson Ltd. (8.5)

Elton, C. 1930. *Animal ecology and evolution.* Oxford: Clarendon (7.5)

Elton, C. 1931. The study of epidemic diseases among wild animals. *J. Hygiene, Camb.* 31: 435–456. (2.7)

Elton, C. 1942. *Voles, mice and lemmings: Problems in population dynamics.* Oxford: Clarendon. (1.2–1.3, 2.1, 2.2, 2.5–2.8, 3.5, 5.3, 5.4, 7.5)

Elton, C. 1954. Research on rodent control by the Bureau of Animal Population,

September 1939 to July 1947. In D. Chitty, ed. *Control of rats and mice.* Vols 1 and 2: Rats, pp. 1-24. Oxford: Clarendon. (4)

Elton, C.S. 1955a. Discussion. In J.B. Cragg and N.W. Pirie, eds. *The numbers of man and animals.* London: Oliver and Boyd, pp. 82–83. (2.4)

Elton, C. 1955b. Review of Andrewartha and Birch (1954). *Nature* 176: 419. (7.4).

Elton, C. 1962. Review of Wynne-Edwards (1962). *Nature* 197: 634. (8.9)

Elton, C. and M. Nicholson. 1942a. Fluctuations in numbers of the muskrat *(Ondatra zibethica)* in Canada. *J. Anim. Ecol.* 11: 96–126. (7.5, 10.1)

Elton, C. and M. Nicholson. 1942b. The ten-year cycle in numbers of the lynx in Canada. *J. Anim. Ecol.* 11: 215–244. (10.1)

Elton, C., E.B. Ford, J.R. Baker, and A.D. Gardner. 1931. The health and parasites of a wild mouse population. *Proc. Zool. Soc. Lond.* 657–721. (1.6, 2.1, 2.7)

Elton, C., D.H.S. Davis, and G.M. Findlay. 1935. An epidemic among voles *(Microtus agrestis)* on the Scottish border in the spring of 1934. *J. Anim. Ecol.* 4: 277–288. (2.2, 2.3, 2.5, 2.7, 5.3, 5.4, 6.2)

Erlinge, S. 1987. Predation and noncyclicity in a microtine population in southern Sweden. *Oikos* 50: 347–352. (2.5)

Erlinge, S, J. Agrell, J. Nelson, and J. Sandell. 1990. Social organization and population dynamics in a *Microtus agrestis* population. In R.H. Tamarin, R.S. Ostfeld, S.R. Pugh, and G. Bujalska, eds. *Social systems and population cycles in voles,* pp. 45–58. Basel: Birkhäuser Verlag, (2.5, 5.4, 5.7, 9.2)

Errington, P.L. 1967. *Of predation and life.* Ames, IA: Iowa State University Press. (7.5)

Evans, F.C. 1942. Studies of a small mammal population in Bagley Wood, Berkshire. *J. Anim. Ecol.* 11: 182–197. (8.8)

Fager, E.W. 1968. The community of invertebrates in decaying oak wood. *J. Anim. Ecol.* 37: 121–142. (7.5)

Feyerabend, P. 1975. In Mackay (1991). (12.1)

Feynman, R.P. 1966. In Mackay (1991). (1.1)

Findlay, G.M. and A.D. Middleton. 1934. Epidemic disease among voles *(Microtus)* with special reference to *Toxoplasma. J. Anim. Ecol.* 3: 150–160. (2.7, 5.7)

Fischlin, A. and W. Baltensweiler. 1979. Systems analysis of the larch budmoth system. Part 1: The larch—larch budmoth relationship. *Mitt. Schweiz. Entomol. Ges.* 52: 273–289. (2.4)

Ford, J., H. Chitty, and A.D. Middleton. 1938. The food of partridge chicks *(Perdix perdix)* in Great Britain. *J. Anim. Ecol.* 7: 251–165. (8.1)

Frank. F. 1954. Beiträge zur Biologie der Feldmaus, *Microtus arvalis* (Pallas). I. Gehegeversuche. *Zool. Jb. (Syst.) Oekol. Geogr. Tiere,* 82: 354–404. (6.5, 11.1)

Franklin, K.J. 1957. *Movement of the heart and blood in animals. An anatomical essay by William Harvey translated from the original Latin [De Motu Cordis].* Oxford: Blackwell. (2.5)

Franklin, K.J. 1958. *The circulation of the blood. Two anatomical essays by William Harvey together with nine lettters written by him. The whole translated from the Latin [De circulatione sanguinis] and slightly annotated.* Oxford: Blackwell. (1.6)

Franz, F. 1953. Über die genetischen Grundlagen des Zusammenbruchs einer Massenvermehrung aus inneren Ursachen. *Z. Angew. Ent.* 31: 228–260. (8.5)

Frenkel, J.K. and S. Friedlander, 1951. Toxoplasmosis: pathology of neonatal disease; Parthenogenesis, diagnosis, and treatment. *Publ. Health Serv. Publ., Wash.,* 141: 1–107. (2.7)

Gaines, M.S. 1981. Importance of genetics to population dynamics. In M.H. Smith and J. Joule, eds. *Mammalian population genetics*. Symp. Am. Soc. Mammal., pp. 1–27. Athens, GA, University of Georgia Press. (1.7, 12.5)

Gaines, M.S. and C.J. Krebs. 1971. Genetic changes in fluctuating vole populations. *Evolution* 25: 702–723. (12.1, 12.4)

Gaines, M.S., L.B. McClenaghan, and R.K. Rose, 1978. Temporal patterns of allozymic variation in fluctuating populations of *Microtus ochrogaster*. *Evolution* 32: 723–739. (11.2)

Gaines, M.S., N.C. Stenseth, M.L. Johnson, R.A. Ims, and S. Bondrup-Nielsen. 1991. A reponse to solving the enigma of population cycles with a multifactorial perspective. *J. Mammal* 72: 627–631. (12.2)

Gardner, M. 1981. *Science: Good, bad and bogus*. Buffalo, NY: Prometheus Books. (P, 11.2)

Getz, L.L., J.E. Hofmann, B.J. Klatt, L. Verner, F.R. Cole, and R.L. Lindroth. 1987. Fourteen years of population fluctuations of *Microtus ochrogaster* and *M. pennsylvanicus* in east central Illinois. *Can. J. Zool.* 65: 1317–1325. (2.2)

Geyl, P. 1955. *Debates with historians*. London: B.T. Batsford Ltd. (2.5, 11.2)

Gilbert, B.S. and C.J. Krebs. 1981. Effects of extra food on *Peromyscus* and *Clethrionomys* populations in the southern Yukon. *Oecologia* 51: 326–321. (2.5)

Goddard, T. Russell. 1935. A census of short-eared owls (*Asio f. flammeus*) at Newcastleton, Roxburghshire, 1935. *J. Anim. Ecol.* 4: 289–290. (2.5)

Godfrey, G.K. 1953. A technique for finding *Microtus* nests. *J. Mammal.* 34: 503–505. (6.2)

Godfrey, G.K. 1954. Tracing field voles (*Microtus agrestis*) with a Geiger-Müller counter. *Ecology* 35: 5–10. (6.2)

Godfrey, G.K. 1955. Observations on the nature of the decline in numbers of two *Microtus* populations. *J. Mammal.* 36: 209–214. (5.7, 6.2)

Gorman, M.R., M.H. Ferkin, R.J. Nelson, and I. Zucker. 1993. Reproductive status influences odor preferences of the meadow vole, *Microtus pennsylvanicus*, in winter day lengths. *Can. J. Zool.* 71: 1748–1754. (12.1)

Green, H.C. (Trans.) 1957. *An introduction to experimental medicine*. [by Claude Bernard], New York: Dover. London: Constable & Co. Ltd. (1.4, 1.5, 2.9, 3.3, 5.6, 5.7, 6.2, 6.4, 8.5, 8.6, 9.4, 11.2)

Green, R.G. and C.A. Evans. 1940. Studies on a population cycle of snowshoe hares on the Lake Alexander Area . . . *J. Wildl. Manag.* 4: 220–238, 267–278, 347–358. (3.4)

Green, R.G. and C.L. Larson. 1938. Shock disease and the snowshoe hare cycle. *Science* 87: 298–299. (3.4)

Green, R.G., C.L. Larson, and J.F. Bell. 1939. Shock disease as the cause of the periodic decimation of the snowshoe hare. *Am. J. Hyg.* 30, Sec. B: 83–102. (3.4)

Griffith, A. S. 1930. *A system of bacteriology*. Vol. 5: 213. London. (2.8)

Gross, A.O. 1947. Cyclic invasions of the snowy owl and the migration of 1945–46. *Auk* 64: 584–601. (1.3)

Gustafsson, T.O., C.B. Andersson, , and L.M. Westlin. 1983a. Reproduction in laboratory colonies of bank vole, *Clethrionomys glareolus*, originating from populations with different degrees of cyclicity. *Oikos* 40: 182–188. (12.5)

Gustafsson, T.O., C.B. Andersson, and N.E.I. Nyholm. 1983b. Comparison of sensitivity to social suppression of sexual maturation in captive male bank voles, *Clethrionomys glareolus*, originating from populations with different degrees of cyclicity. *Oikos* 41: 250-254. (12.5)

Hagen, J.B. 1992. Review of Crowcroft (1991). *J. Hist. Biol.* 25: 171–173. (Ack)

Hairston, N.G. 1989. *Ecological experiments: Purpose, design, and execution.* Cambridge: Cambridge University Press. (1.5)

Hall, T.H. 1969. *Sherlock Holmes: Ten literary studies.* London: Gerald Duckworth. (11.2)

Hallam, A. 1983. *Great Geological Controversies.* Oxford: Oxford University Press. (1.4, 10.3)

Halle, S. and U. Lehmann. 1992. Cycle-correlated changes in the activity behaviour of field voles, *Microtus agrestis. Oikos* 64: 489–497. (8.7, 12.5)

Hamilton, W.D. 1970. Selfish and spiteful behaviour in an evolutionary model. *Nature* 228: 1218–1220. (7.3, 12.4)

Hamilton, W.J., Jr. 1937. The biology of microtine cycles. *J. Agric. Res.* 54: 779–790 (5.4, 5.7)

Hanski, I., L. Hansson, and H. Henttonen. 1991. Specialist predators, generalist predators, and the microtine rodent cycle. *J. Anim. Ecol.* 60: 353–367. (2.5, 10.1, 12.2)

Hanski, I., P. Turchin, E. Korpimäki, and H. Henttonen. 1993. Population oscillations of boreal rodents: Regulation by mustelid predators leads to chaos. *Nature* 364: 232–235. (5.7)

Hansson, L. 1984. Composition of cyclic and noncyclic vole populations: On the causes of variation in individual quality among *Clethrionomys glareolus* in Sweden. *Oecologia* 63: 199–206. (12.5)

Hansson, L. 1987. An interpertation of rodent dynamics as due to trophic interactions. *Oikos* 50: 308-318. (1.7)

Hansson, L. 1988. Parent-offspring correlations for growth and reproduction in the vole *Clethrionomus* [sic] *glareolus* in relation to the Chitty Hypothesis. *Z. Säugetierkunde* 53: 7–10. (1.7)

Hansson, L. and H. Henttonen. 1988. Rodent dynamics as community processes. *Trends Ecol. Evol.* 3: 195–200. (10.3, 12.2)

Hansson, L. and N.C. Stenseth. eds. 1988. Modelling small rodent population dynamics. *Oikos* 52: 138–229. (1.4)

Hardy, W.B. 1985. In Mckay (1991). (4)

Haukioja, E., K. Kapiainen, P. Niemelä, and J. Tuomi. 1983. Plant availability hypothesis and other explanations of herbivore cycles: Complementary or exclusive alternatives? *Oikos* 40: 419–432. (3.1)

Hawking, S.W. 1988. *A brief history of time: From the big bang to black holes.* Toronto: Bantam Books. (8.6, 9.3)

Hempel, C.G. 1966. *Philosophy of natural science.* Englewood Cliffs, NJ: Prentice-Hall. (3.1, 8.2)

Henttonen, H. and A. Kaikusalo, 1993. Lemming movements. In Stenseth and Imms (1993), q.v., pp. 157–186. (1.2)

Henttonen, H., T. Oksanen, A. Jortikka, and V. Haukisalmi. 1987. How much do weasels shape microtine cycles in the northern Fennoscandian taiga? *Oikos* 50: 353–365. (1.7, 5.4, 6.2, 7.3, 9.2, 9.3)

Heske, E.J. and S. Bondrup-Nielsen. 1990. Why spacing behavior does not stabilize density in cyclic populations of microtine rodents. *Oecologia* 83: 91–98. (10.3)

Heske, E.J., R.S. Ostfeld, and W.Z. Lidicker, Jr. 1987. Does social behavior drive vole cycles? An evaluation of competing models as they pertain to California voles. *Can. J. Zool.* 66: 1153–1159. (3.4)

Heske, E.J., R.A. Ims, and H. Steen. 1993. Four experiments on a Norwegian subalpine microtine rodent assemblage during a summer decline. In Stenseth and Ims (1993), q.v., pp. 411–424. (9.3)

Hestbeck, J.B. 1982. Population regulation of cyclic mammals: The social fence hypothesis. *Oikos* 39: 157–163. (12.5)

Hik, D.S. 1995. Does risk of predation influence population dynamics? Evidence from the cyclic decline of snowshoe hares. *Wildl. Res.* 22: 115-129. (2.5, 5.3, 9.2, 10.1)

Hilborn, R. and S.C. Stearns. 1982. On inference in ecology and evolutionary biology: The problem of multiple causes. *Acta Biotheoret.* 31: 145–164. (12.2)

Holsinger, K.E. 1981. Comment: The blunting of Occam's Razor, or to hell with parsimony. *Can. J. Zool.* 59: 144–146. (1.1, 1.6)

Hospers, J. 1953. *An introduction to philosophical analysis.* Englewood Cliffs, NJ: Prentice-Hall. (5.7, 11.2)

Hudson, P.J. and A.P. Dobson. 1990. Red grouse population cycles and the population dynamics of the caecal nematode *Trichostrongylus tenuis.* In A.N. Lance and J.H. Lawton, eds. *Red grouse population processes. Proceedings of a workshop convened by the British Ecological Society and the Royal Society for the protection of Birds,* pp. 5–19. (2.7)

Hull, D. (1973). *Darwin and his critics. The reception of Darwin's theory of evolution by the scientific community.* Cambridge: Harvard University Press. (1.5, 8.6)

Hull, D. 1988. *Science as a process. An evolutionary account of the social and conceptual development of science.* Chicago: University of Chicago Press. (P, 1.1, 1.5, 1.6, 7.1, 7.3, 8.6, 11.2, 12.1)

Hurlbert, S.H. 1984. Pseudoreplication and the design of ecological field experiments. *Ecol. Monog.* 54: 187– 211. (2.5)

Huxley, L. (ed.) 1903. *Life and letters of Thomas Henry Huxley.* [LL] London: Macmillan, 3 Vols. (3.1, 8.6, 11.1)

Huxley, T.H. 1893–1906. *Collected Essays.* [CE]. London: Macmillan 9 Vols. (P, 7.4, 8.9, 11.1)

Jarvinen, A. 1987. Microtine cycles and plant production: What is cause and effect? *Oikos* 49: 352–357. (12.2)

Jenkins, D. 1956. Factors governing population density in the partridge. Unpubl. D. Phil. thesis, Oxford. (8.2)

Jenkins, D. 1961a. Population control in protected partridges (*Perdix perdix*). *J. Anim. Ecol.* 30: 235–258. (8.2)

Jenkins, D. 1961b. Social behaviour in the partridge *Perdix perdix. Ibis* 103a:155– 188. (8.2)

Jenkins, D. and Watson, A. 1967. Population control in red grouse and rock ptarmigan in Scotland. *Finnish Trans. Congr. Int. Union Game Biol.,* 8: 121–141. (8.2)

Jenkins, D., A. Watson, and G.R. Miller. 1967. Population fluctuations in the red grouse *Lagopus lagopus scoticus. J. Anim. Ecol.* 36: 97–122. (1.7, 8.2)

Johnson, A.H., ed. 1947. *The wit and wisdom of Whitehead.* Boston: Beacon Press. (1.1, 5.3)

Judson, H.F. 1979. *The eighth day of creation: Makers of the revolution in biology.* New York: Touchstone Books. (P, Ack, 3.1, 3.5, 5.1, 5.2, 6.6, 7.1, 7.2, 7.3, 10.1)

Keith, L.B. 1963. *Wildlife's ten-year cycle.* Madison: University of Wisconsin Press. (10.1, 10.2)

Keith, L.B. 1974. Some features of population dynamics in mammals. *Proc. Int. Congr. Game Biologists* 11: 17–58. (3.1)

Keith, L.B. 1990. Dynamics of snowshoe hare populations. In H.H. Genoways, ed. *Current mammalogy,* Vol. 2., pp. 119–195. New York: Plenum. (5.3, 9.4)

Keith, L.B., and L.A. Windberg. 1978. A demographic analysis of the snowshoe hare cycle. *Wildl. Monogr.* 58: 1–70. (1.7, 5.3, 5.8)

Keith, L.B., J.R. Cary, O.J. Rongstad, and M.C. Brittingham. 1984. Demography and ecology of a declining snowshoe hare population. *Wildl. Monogr.* 90: 1–43. (3.1)

Kidd, N.A.C. and G.B. Lewis. 1987. Can vertebrate predators regulate their prey? A reply. *Am. Nat.* 130: 448–453. (2.5)

Kikkawa, J. 1964. Movement, activity and distribution of the small rodents *Clethrionomys glareolus* and *Apodemus sylvaticus* in woodland. *J. Anim. Ecol.* 33: 259–299. (12.4)

Kingsland, S.E. 1985. *Modeling nature. Episodes in the history of population ecology.* Chicago: University of Chicago Press. (1.4)

Kitaigorodskii, A.I. 1975. In Mackay (1991). (11.2)

Kneale, W. 1949. *Probability and induction.* Oxford: Clarendon. (1.5)

Koponen, T., A. Kokkonen, and O. Kalela, 1961. On a case of spring migration in the Norwegian lemming. *Ann. Acad. Sci. Fenn.* Ser.A. IV. 52: 3–30. (1.2)

Krebs, C.J. 1964a. The lemming cycle at Baker Lake, Northwest Territories, during 1959–62. *Arctic Inst. North Am.* Tech. Paper No. 15: 1–104. (1.2, 5.2, 5.3, 5.8, 9.4)

Krebs, C.J. 1964b. Cyclic variation in skull-body regressions of lemmings. *Can. J. Zool.* 42: 631–643. (9.4)

Krebs, C.J. 1966. Demographic changes in fluctuating populations of *Microtus californicus. Ecol. Monogr.* 36: 239–273. (8.8, 9.3, 9.4, 11.1).

Krebs, C.J. 1970. Genetic and behavioral studies on fluctuating vole populations. *Proc. Adv. Study Inst. Dynamics Numbers Popul. (Osterbeek, 1970):* 243–256. (12.4)

Krebs, C.J. 1978. A review of the Chitty Hypothesis of population regulation. *Can. J. Zool.* 56: 2463–2480. (9.3)

Krebs, C.J. 1979. Dispersal, spacing behaviour, and genetics in relation to population fluctuations in the vole *Microtus townsendii. Fortschr. Zool.* 25: 61–77. (3.3, 5.2, 5.4, 12.4, 12.5)

Krebs, C.J. 1985. *Ecology: The experimental analysis of distribution and abundance.* 3rd ed. New York: Harper & Row. (1.1, 2.5, 12.5)

Krebs, C.J. 1988. The experimental approach to rodent population dynamics. *Oikos* 88: 143–149. (1.4, 12.5)

Krebs, C.J. 1992a. Review of Crowcroft (1991). *Science* 252: 1010. (Ack)

Krebs, C.J. 1992b. The role of dispersal in cyclic rodent populations. In N.C. Stenseth and W.Z. Lidicker, Jr., eds. *Animal dispersal: Small mammals as a model,* pp. 160-175. London: Chapman and Hall. (9.3, 11.1, 12.5)

Krebs, C.J. 1993. Are lemmings large *Microtus* or small reindeer? A review of lemming cycles after 25 years and recommendations for future work. In Stenseth and Ims, eds. *The biology of lemmings.* Symp. Linn. Soc. Ser. 15, pp. 247–260. (12.5)

Krebs, C.J. (1995a). Population regulation. *Encyclopedia of Environmental Biology,* Vol. 3. San Diego, CA: Academic Press, pp. 183–202. (7.5)

Krebs, C.J. (1995b). Two paradigms of population regulation. *Wildl. Res.* 22: 1–10. (7.5)

Krebs, C.J., S. Boutin, R. Boonstra, A.R.E. Sinclair, J.N.M. Smith, M.R.T. Dale, K.

Martin, and R. Turkington. 1995. Impact of food and predation on the snowshoe hare cycle. *Science* 269: 1112–1115. (2.5)

Krebs, C.J., S. Boutin, and B.S. Gilbert. 1986a. A natural feeding experiment on a declining snowshoe hare population. *Oecologia* 70: 194–1197. (9.2)

Krebs, C.J., D. Chitty, G. Singleton, and R. Boonstra 1995. Can changes in social behaviour help to explain house mouse plagues in Australia? *Oikos* (73: 429–434) (12.5)

Krebs, C.J. and K.T. DeLong. 1965. A *Microtus* population with supplemental food. *J. Mammal.* 46: 566–573. (9.2, 9.3)

Krebs , C.J., B.S. Gilbert, S. Boutin, A.R.E. Sinclair, and J.N.M. Smith. 1986b. Population biology of snowshoe hares. I. Demography of food-supplemented populations in the southern Yukon, 1976–84. *J. Anim. Ecol.* 55: 963–982. (9.2)

Krebs, C.J., B.L. Keller, and R.H. Tamarin. 1969. *Microtus* population biology: demographic changes in fluctuating populations of *M. ochrogaster* and *M. pennsylvanicus* in southern Indiana. *Ecology* 50: 587–607. (6.3, 12.4)

Krebs, C.J. and Myers, J.H. 1974. Population cycles in small mammals. *Adv. Ecol. Res.* 8: 267–399.

Krebs, C.J., I. Wingate, J. LeDuc, J.A. Redfield, M. Taitt, and R. Hilborn. 1976. *Microtus* population biology: Dispersal in fluctuating populations of *M. townsendii*. *Can. J. Zool.* 54: 79–95. (5.8)

Krebs, J.R. 1992. Review of Crowcroft (1991). *Trends Ecol. Evol.* 6: 302–303. (Ack)

Krebs, J.R. and N.B. Davies (1993). *An introduction to behavioural ecology*. 3rd ed. Oxford: Blackwell. (8.9)

Kuhn, T.S. 1962. *The structure of scientific revolutions*. Chicago: University of Chicago Press. (3.4, 7.3, 7.5)

Lack, D. 1954a. *The natural regulation of animal numbers*. Oxford: Clarendon. (7.3, 8.1)

Lack, D. 1954b. Cyclic mortality. *J. Wildl. Manag.* 18: 25–37. (7.3)

Lack, D. 1954c. Letter to the editor. *J. Anim. Ecol.* 23: 381. (8.1)

Lack, D. 1955. The mortality factors affecting adult numbers. In J.B. Cragg and N.W. Pirie, eds. *The numbers of man and animals*. London: Oliver and Boyd, pp. 47–55. (7.3)

Lack, D. 1964. A long-term study of the great tit *(Parus major)*. *J. Anim. Ecol.* 33 (Jubilee Suppl.): 159–173. (1.5)

Lack, D. 1965. *Enjoying ornithology*. London: Methuen & Co. Ltd. (4, 7.3)

Lack, D. 1966. *Population studies of birds*. Oxford: Clarendon. (1.5, 7.3, 8.9)

Laine, K. and H. Henttonen. 1983. The role of plant production in microtine cycles in northern Fennoscandia. *Oikos* 40: 407–418. (12.2)

Lakatos, I. 1970. Falsification and the methodology of scientific research programmes. In I. Lakatos and A. Musgrave, eds. *Criticism and the growth of knowledge*, pp. 91–196. Cambridge: Cambidge University Press, (6.1, 11.2, 12.5)

Lambin, X. and C.J. Krebs, 1991. Spatial organization and mating system of *Microtus townsendii*. *Behav. Ecol. Sociobiol.* 28: 353–363. (6.2, 11.1, 12.5).

Lambin, X. and C.J. Krebs. 1993. Influence of female relatedness on the demography of Townsend's vole populations in spring. *J. Anim. Ecol.* 62: 536-550. (6.2, 12.5)

Lance, A. N. and J.H. Lawton, eds. 1990. *Red grouse population processes. Proceedings of a workshop convened by The British Ecological Society and The Royal Society for the Protection of Birds*. (8.2)

Lawton, J.H. 1992. (Modest) advice for graduate students. *Oikos* 65: 361–362. (1.7)

Leacock, S. 1961. *My discovery of England.* Toronto: McClelland and Stewart. (3.1).

LeDuc, J. and C.J. Krebs. 1975. Demographic consequences of artificial selection at the LAP locus in voles *(Microtus townsendii). Can. J. Zool.* 53: 1825–1840. (12.1, 12.4).

Leslie, P.H. 1946. Population theories. A review of *La Lotta per l'Esistenza,* by U. d'Ancona. *J. Anim. Ecol.* 15: 107. (3.3)

Leslie, P.H. 1959. The properties of a certain lag type of population growth and the influence of an external random factor on a number of such populations. *Physiol. Zoöl.* 32: 151–159. (10.3)

Leslie, P.H. and D. Chitty. 1951. The estimation of population parameters from data obtained by means of the capture-recapture method. I. The maximum-likelihood equations for estimating the death-rate. *Biometrika* 38: 269–292. (5.4)

Leslie, P.H., D. Chitty, and H. Chitty. 1953. The estimation of population parameters from data obtained by means of the capture-recapture method. III. An example of the practical applications of the method. *Biometrika* 40: 137–169. (5.1, 5.4)

Leslie, P.H. and R.M. Ranson. 1940. The mortality, fertility and rate of natural increase of the vole *(Microtus agrestis)* as observed in the laboratory. *J. Anim. Ecol.* 9: 27–52. (3.3)

Leslie, P.H., J.S. Tener, M. Vizozo, and H. Chitty. 1955. The longevity and fertility of the Orkney vole, *Microtus orcadensis,* as observed in the laboratory. *Proc. Zool. Soc. Lond.* 125: 115–125. (1.3, 5.6, 8.6)

Levin, S.A. 1981. The role of theoretical ecology in the description and understanding of populations in heterogeneous environments. *Am. Zool.* 21: 865–875. (1.4)

Lidicker, W.J., Jr. 1975. The role of dispersal in the demography of small mammals. In K. Petrusewicz, F.B. Golley, and L. Ryszkowski, eds. *Small mammals: Their productivity and population dynamics,* pp. 103–128. London: Cambridge University Press. (1.7, 9.3)

Lidicker, W.Z., Jr. 1988. Solving the enigma of microtine "cycles". *J. Mammal.* 69: 225–235. (3.5, 12.2)

Lidicker, W.Z., Jr. 1991. In defense of a multifactor perspective in population ecology. *J. Mammal.* 72: 631–635. (12.2)

Lidicker, W.Z., Jr. 1994. Population ecology. In E.C. Birney and J.R. Choate, eds. *Seventy-five years of mammalogy (1919–1994),* pp. 323–347. Spec. Publ. Am. Soc. Mammal.No.11. (12.5)

Lidicker, W.Z., Jr. and R.S. Ostfeld. 1991. Extra-large body size in California voles: Causes and fitness consequences. *Oikos* 61: 108–121. (5.8)

Livingstone, R.W. 1952. *Education and the spirit of the age.* Oxford: Clarendon. (11.1)

MacArthur, R.H. and E.O. Wilson. 1967. *Theory of island Biogeography.* Princeton, NJ: Princeton University Press. (11.1)

Macfadyen, A. 1992. Obituary: Charles Sutherland Elton. *J. Anim. Ecol.* 61: 499–502. (Ack)

Mackay, A.L. 1991. *A dictionary of scientific quotations.* Bristol: Adam Hilger. (1.1, 1.5, 3.2, 4, 8.5, 11.2, 12.1)

McIntosh, R.P. 1985. *The background of ecology: Concept and theory.* Cambridge: Cambridge University Press. (1.1)

MacLulich, D.A. 1937. Fluctuations in the numbers of the varying hare *(Lepus americanus). Univ. Toronto Stud. Biol.* Ser. No. 43: 1–136. (3.1, 10.1, 10.2)

Magee, B. 1973. Popper. London: Fontana/Collins. (11.2)

Malcolm, J.R. and R.J. Brooks. 1993. Cyclic variation in skull-body regressions of collared lemmings: Differential representation of seasonal cohorts. In Stenseth and Ims, eds. *The biology of lemmings.* Symp. Linn. Soc. Ser. 15, pp. 135–155. (9.4)

Marsden, W. 1964. *The lemming year.* London: Chatto and Windus, (1.2).

Mather, K. 1961. Competition and co-operation. *Symp. Soc. Exp. Biol.* 15: 264–281. (11.1).

May, R.M. 1981. Models for single populations. In R.M. May, ed. *Theoretical Ecology: Principles and applications.* 2nd ed. Sunderland, MA: Sinauer Associates Inc., pp. 5–29. (1.3, 2.5)

May, R.M. and R.M. Anderson. 1978. Regulations and stability of host-parasite population interactions. II. Destabilizing processes. *J. Anim. Ecol.* 47: 249-267. (2.7)

Maynard Smith. 1964. Group selection and kin selection. *Nature* 201: 1145–1147. (8.9)

Medawar, P.B. 1957. *The uniqueness of the individual.* London: Methuen and Co. Ltd. (2.3, 2.9, 5.6, 7.4, 11.1, 11.2)

Medawar, P.B. 1964. Is the scientific paper a fraud? In D. Edge, ed.: *Experiment: A series of scientific case histories.* London: BBC. pp. 7–12. [Reprinted in Medawar (1990)]. (1.5)

Medawar, P.B. 1967. *The art of the soluble.* London: Methuen and Co. Ltd. [Reprinted in part in Medawar (1984a).] (1.5, 4, 7.2, 8.5, 11.1, 11.2)

Medawar, P.B. 1969. Induction and intuition in scientific thought, *Mem. Am. Philos. Soc.* 75. [Reprinted in Medawar (1984).] (2.9, 6.3, 8.6, 10.3)

Medawar, P.B. 1979. *Advice to a young scientist.* New York: Harper & Row. (Ack, 1.1, 5.1, 12.2, 12.5)

Medawar, P. 1984a. *Pluto's republic.* Oxford: Oxford University Press. (Ack, 1.1, 8.3)

Medawar, P.B. 1984b. *The limits of science.* New York: Harper & Row. (1.1, 8.6, 11.2)

Medawar, P. 1986. *Memoir of a thinking radish: An autobiography.* Oxford: Oxford University Press. (1.1, 2.3)

Medawar, P. 1990. The threat and the glory: Reflections on science and scientists. New York: Harper Collins. (4, 11.2)

Medawar, P. 1991. Foreword to the first edition of Mackay (1991). (4)

Middleton, A.D. 1930. Cycles in the numbers of British voles (*Microtus*). *J. Ecol.* 18: 156–165. (2.1)

Middleton, A.D. 1931. A further contribution to the study of cycles in British voles (*Microtus*). *J. Ecol.* 19: 190–199. (2.1)

Middleton, A.D. 1934. Periodic fluctuations in British game populations. *J. Anim. Ecol.* 3: 231–249. (8.1)

Middleton, A.D. 1935a. Factors controlling the population of the partridge *(Perdix perdix)* in Great Britain. *Proc. Zool. Soc. Lond.* 4: 795–815. (8.1)

Middleton, A.D. 1935b. The population of partridges (*Perdix perdix)* in 1933 and 1934 in Great Britain. *J. Anim. Ecol.* 4: 137–145. (8.1)

Mihok, S. 1981. Chitty's hypothesis and behaviour in subarctic red-backed voles *Clethrionomys gapperi. Oikos* 36: 281–295. (11.1)

Mihok, S. 1988. The role of mortality in models of small rodent population dynamics. *Oikos* 52: 215–218. (5.8, 6.2)

Mihok, S. and R. Boonstra. 1992. Breeding performance in captivity of meadow voles (*Microtus pennsylvanicus*) from decline- and increase-phase populations. *Can. J. Zool.* 70: 1561–1566. (5.6, 8.8, 12.1)

Mihok, S. and W.A. Fuller. 1981. Morphometric variation in *Clethrionomys gapperi:* Are all voles created equal? *Can. J. Zool.* 59: 2275–2283. (5.3)

Mihok, S., B.N. Turner, and S.L Iverson. 1985. The characterization of vole population dynamics. *Ecol. Monog.* 55: 399–420. (3.3, 5.8, 12.5)

Mitter, C. and J.C. Schneider. 1987. Genetic change and insect outbreaks. In P. Barbosa and J.C. Schultz, eds. *Insect outbreaks* pp. 505–532 London: Academic Press. (12.1)

Moen, J. 1990. Summer grazing by voles and lemmings upon subarctic snow-bed and tall herb meadow vegetation—an enclosure experiment. *Holarctic Ecol.* 13: 316–324. (9.2)

Moran, P.A.P. 1952. The statistical analysis of game-bird records, I. *J. Anim. Ecol.* 21: 154–158. (8.1)

Moran, P.A.P. 1953a. The statistical analysis of the Canadian lynx cycle. I. Structure and prediction. *Aust. J. Zool.* 1: 164–173. (10.1)

Moran, P.A.P. 1953b. The statistical analysis of the Canadian lynx cycle II. Synchronization and meteorology. *J. Anim. Ecol.* 1: 291–298. (10.1, 10.3)

Moran, P.A.P. 1954. The statistical analysis of game-bird records. II. *J. Anim. Ecol.* 23: 35–37. (8.1).

Morris, R.F. 1963. Foliage depletion and the spruce budworm. In R.M. Morris, ed. *The dynamics of epidemic spruce budworm populations.* Mem. Ent. Soc. Can. No. 31, pp. 223-228 and Plate 1.1 (2.4)

Morris, R.F. 1971. Observed and simulated changes in genetic quality in natural populations of *Hyphantria cunea. Can. Ent.* 103: 893–906. (12.1)

Moss, R., P. Rothery, and I.B.Trenholm. 1985. The inheritance of social dominance rank in red grouse (*Lagopus lagopus scoticus*). *Aggress. Behav.* 11: 253–259. (12.1)

Moss, R. and A. Watson. 1990. Predicting, manipulating and understanding red grouse population fluctuations. In A.N. Lance and J.H. Lawton, eds. *Red grouse population processes. Proceedings of a workshop convened by The British Ecological Society and the Royal Society for the Protection of Birds,* pp. 72-83. (9.3)

Moss, R. and A. Watson. 1991. Population cycles and kin selection in red grouse *Lagopus lagopus scoticus. Ibis* 133 Suppl. 1: 113–120. (2.5, 9.3, 10.1, 11.1, 12.5).

Moss, R., A. Watson, and P. Rothery. 1984. Inherent changes in the body size, viability and behaviour of a fluctuating red grouse (*Lagopus lagopus scoticus*) population. *J. Anim. Ecol.* 53: 171–189. (12.5).

Myers, J.H. 1990. Population cycles of western tent caterpillars: Experimental introductions and synchrony of fluctuations. *Ecology* 71: 986–995. (8.2, 9.3)

Myers, J.H. 1993. Population outbreaks in forest Lepidoptera. *Am. Sci.* 81: 240–251. (8.2)

Myers, J.H. and C.J. Krebs. 1971. Genetic, behavioral, and reproductive attributes of dispersing field voles *Microtus pennsylvanicus* and *Microtus ochrogaster. Ecol. Monogr.* 41: 53–78. (1.7, 5.8).

Myers, J.H. and L.D. Rothman. 1995. Field experiments to study regulation of fluctuating populations. In N. Cappuccino and P. Price, eds. *Population Dynamics: New Approaches and Synthesis,* pp. 229–250 San Diego, CA: Academic Press, (8.2, 9.3)

Myllymäki, A. 1977. Demographic mechanisms in the fluctuating populations of the field vole *Microtus agrestis. Oikos* 29: 468–493. (5.2)

Myllymäki, A, J, Aho, E.A. Lind, , and J. Tast, 1962. Behaviour and daily activity of the Norwegian lemming, *Lemmus lemmus* (L.), during autumn migration. *Ann. Zool. Soc. 'Vanamo':* 24 (2): 1–31. (1.2)

Nelson, R.J. 1987. Photoperiod-nonresponsive morphs: A possible variable in microtine population-density fluctuations. *Am. Nat.* 130: 350–369. (12.1)

Nelson, J., J. Agrell, S. Erlinge, and M. Sandell. (1991). Reproduction of different female age categories and dynamics in a non-cyclic field vole, *Microtus agrestis,* population. *Oikos* 61: 73–78. (5.8)

Newsome, A.E. 1967. A simple biological method of measuring the food supply of house-mice. *J. Anim. Ecol.* 36: 645–650. (9.2)

Newsome, A.E. 1969. A population study of house-mice temporarily inhabiting a South Australian wheatfield. *J. Anim. Ecol.* 38: 341–359. (9.2)

Newson, J. 1962. Seasonal differences in reticulocycte count, haemoglobin level and spleen weight in wild voles. *Br. J. Haematol.* 8: 296–302. (8.5)

Newson, J. and D. Chitty. 1962. Haemoglobin levels, growth and survival in two *Microtus* populations. *Ecology* 43: 733–738. (8.8)

Newson, R. 1963. Differences in numbers, reproduction and survival between two neighboring populations of bank voles (*Clethrionomys glareolus*). *Ecology* 44: 110–120. (9.1)

Nygren, J. 1978. Interindividual influence on diurnal rhythms of activity in cycling and noncycling populations of the field vole (*Microtus agrestis* L.). *Oecologia* 35: 231–239. (8.7, 12.5)

Nygren, J. 1980. Allozyme variation in natural populations of field vole (*Microtus agrestis* L.). III. survey of a cyclically density-varying population. *Hereditas* 93: 125–135. See also *Hereditas* 92: 65–72; 93: 107–114. (12.5)

Nyholm, N.E. and P. Meurling. 1979. Reproduction of the bank vole, *Clethrionomys glareolus,* in northern and southern Sweden during several seasons and in different phases of the vole population cycle. *Holarctic Ecol.* 2: 12–20. (12.5)

Oksanen, L. and T. Oksanen. 1992. Long-term microtine dynamics in north Fennoscandian tundra: The vole cycle and the lemming chaos. *Ecography* 15: 226–236. (11.2)

Oreskes, N., K. Shrader-Frechette, and K. Belitz. 1994. Verification, validation, and confirmation of numerical models in the earth sciences. *Science* 263: 641–646. (12.5)

Ostfeld, R.S. 1985. Experimental analysis of aggression and spacing behavior in the California vole. *Can. J. Zool.* 63: 2277–2282. (1.7)

Ostfeld, R.S. 1994. The fence effect reconsidered. *Oikos* 70: 340-348. (12.4, 12.5)

Page, R.E. and A.T. Bergerud. 1984. A genetic explanation for ten-year cycles of grouse. *Oecologia* 64: 54-60. (10.3, 12.5)

Paine, R.T. 1994 *Marine rocky shores and community ecology: An experimentalist's perspective.* The Ecology Institute, Nordburite 23, D-21385 Oldendorf/Luke, Germany. (1.5)

Park, T. 1935. Studies in population physiology. IV. Some physiological effects of conditioned flour upon *Tribolium confusum* and its populations. *Physiol. Zoöl.* 8: 91–115. (5.5)

Parker, G.A. and N. Knowlton. 1980. The evolution of territory size—Some ESS models. *J. Theor. Biol.* 84: 445–476. (7.3)

Pease, J.L., R.H. Vowles, and L.B. Keith, 1979. Interaction of snowshoe hares and woody vegetation. *J. Wildl. Manag.* 43: 43–60. (5.3)

Peters, R.H. 1991. *A critique for ecology.* Cambridge: Cambridge University Press. (1.4)

Philip, C.B. 1939. A parasitological reconnaissance in Alaska with particular reference to varying hares. *J. Mammal.* 20: 82–86. (10.2)

Pimentel, D. 1958. Population regulation and genetic feedback. *Science* 159: 1432–1437. (9.2)

Pitelka, F.A. 1958. Some aspects of population structure in the short-term cycle of the brown lemming in northern Alaska. *Cold Spring Harb. Symp. Quant. Biol.* 22 (1957): 237–251. (12.3)

Pitelka, F.A. 1964. The nutrient-recovery hypothesis for arctic microtine cycles. I. Introduction. In D.J. Crisp, ed. *Grazing in terrestrial and marine environments. A symposium of the British Ecological Society.* pp. 55–56. Oxford: Blackwell, (12.3)

Pitelka, F.A. 1973. Cyclic pattern in lemming populations near Barrow, Alaska. In M.E. Britton, ed.: *Alaskan Arctic Tundra.* Arctic Inst. N. America, Technical Paper No. 24: 199–215. (3.1, 5.3, 12.3)

Platt, J.R. 1964. Strong inference. *Science* 146: 347–353. (2.9, 12.5)

Popper. K.R. 1959. The logic of scientific discovery. London: Hutchinson. (2.9, 11.2, 12.3)

Popper, K.R. 1963. Conjectures and refutations: The growth of scientific knowledge. London: Routledge and Kegan Paul. (P, 1.2, 2.5, 11.2, 12.3).

Potts, G.R., S.C. Tapper, and P.J. Hudson. 1984. Population fluctuations in red grouse: Analysis of bag records and a simulation model. *J. Anim. Ecol.* 53: 21–36. (1.3, 10.1).

Ranson, R.M. 1934. The field vole (*Microtus*) as a laboratory animal. *J. Anim. Ecol.* 3: 70–76. (8.6)

Rasmuson, B., M. Rasmuson, and J. Nygren. 1977. Genetically controlled differences in behaviour between cycling and non-cycling populations of field vole (*Microtus agrestis*). *Hereditas* 87: 33–43. (12.5)

Reichenbach, J. 1959. *The rise of scientific philosophy.* Berkeley: University of California Press. (1.4)

Richards, P.I. 1967. Writers: Fancies and foibles. *Science* 158: 319. (1.6)

Roland, J. 1988. Decline in winter moth populations in North America: Direct versus indirect effect of introduced parasites. *J. Anim. Ecol.* 57: 523–531. (11.1)

Roland, J. 1990. Interaction of parasitism and predation in the decline of winter moth in Canada. In A.D. Watt, S.R. Leather, M.D. Hunter, and N.A.C. Kidd, eds. *Population dynamics of forest insects,* pp. 289–302. Andover: Intercept Ltd. (11.1)

Roland, J. 1994. After the decline: What maintains low winter moth density after successful control? *J. Anim. Ecol.* 63: 392–398. (11.1)

Rose, R.K. and M.S. Gaines. 1981. Relationships of genotype, reproduction, and wounding in Kansas prairie voles. In M.H. Smith and J. Joule, eds. *Mammalian population genetics. Symp. Am. Soc. Mammal.* pp. 160–179. Athens, GA: University of Georgia Press. (12.5)

Rossiter, M.C. 1992. The impact of resource variation on population quality in herbivorous insects: A critical aspect of population dynamics. In M.D. Hunter, T. Ohgushi, and P.W. Price, eds. *Effects of resource distribution on animal-plant interactions.* pp. 13–42. San Diego, CA: Adademic Press. (11.1)

Rossiter, M.C. 1994. Maternal effects hypothesis of herbivore outbreak. *BioScience.* 44: 752–763. (11.1)

Rothstein, S.I. 1979. Gene frequencies and selection for inhibitory traits, with special emphasis on the adaptiveness of territoriality. *Am. Nat.* 113: 317–321. (7.3)

Russell, B. 1931. *The scientific outlook*. London: Allen & Unwin. (9.1, 11.2)

Sandell, M., M. Åström, O. Altegrim, K. Danell, L. Edenius, J. Hjälten, P. Lundberg, T. Palo, R. Pettersson, and G. Sjöberg, G. 1991. "Cyclic" and "non-cyclic" small mammal populations: An artificial dichotomy. *Oikos* 61: 281–284. (5.8, 12.5)

Saucy, F. 1988. Dynamique de population, dispersion et organisation sociale de la forme fouisseuse du campagnol terrestre (*Arvicola terrestris* Scherman (Shaw), Mammalia, Rodentia). Université de Neuchatel. (5.3)

Schaffer, W.M. and R.H. Tamarin, 1973. Changing reproductive rates and population cycles in lemmings and voles. *Evolution* 327: 111–124. (1.7)

Schrödinger, E. 1954. *Nature and the Greeks*. Cambridge: Cambridge University Press. (P, 5.5, 8.4, 9.2)

Schultz, A.M. 1964. The nutrient-recovery hypothesis for arctic microtine cycles. II. Ecosystem variables in relation to arctic microtine cycles. In D.J. Crisp, ed. *Grazing in terrestrial and marine environments. A symposium of the British Ecological Society*. pp. 57–68. Oxford: Blackwell. (12.3)

Schwerdtfeger, F. 1941. Über die Ursachen des Massenwechsels der Insekten. *Z. Angew. Ent.* 28: 254-303. (11.1)

Scott, J.C. 1955. Stress factor in the disc syndrome. *J. Bone Joint Surg. Brit. Number* 37B: 107–111. (6.4)

Selye, H. 1936. A syndrome produced by diverse nocuous agents. *Nature*. 138: 32. (6.1)

Selye, H. 1946. The general adaptation-syndrome and diseases of adaptation. *J. Clin. Endocrinol. Metab.* 6: 217–230. (6.1)

Selye, H. 1950. The physiology and pathology of exposure to stress. Montreal: Acta, Inc. (6.1)

Semeonoff, R. and F.W. Robertson. 1968. A biochemical and ecological study of plasma esterase polymorphism in natural populations of the field vole, *Microtus agrestis* L. *Biochem Genet.* 1:205–227. (12.1, 12.4)

Sheail, J. 1987. *Seventy-five years in ecology: The British Ecological Society*. Oxford: Blackwell. (2.4, 7.5)

Shelford, V.E. 1943. The abundance of the collared lemming (*Dicrostonyx groelandicus* (Tr.) var. *Richardsoni* Mer.) in the Churchill area, 1929 to 1940. *Ecology* 24: 472–484. (1.3)

Shelford, V.E. 1945. The relation of snowy owl migration to the abundance of the collared lemming. *Auk* 62: 592–596. (1.3)

Shreffler, P.A., ed. 1984. *The Baker Street Reader: Cornerstone writings about Sherlock Holmes*. Westport, CT. Greenwood Press. (11.2)

Simberloff, D. 1983. Competition theory, hypothesis-testing, and other commuity ecological buzzwords. *Am. Nat.* 122: 626–635. (P)

Sinclair, A.R.E. 1989. Population regulation in animals. In J.M. Cherrett, ed. *Ecological concepts. The contribution of ecology to an understanding of the natural world. Symp. Br. Ecol. Soc. 29*, pp. 197–241. Oxford: Blackwell. (1.7)

Sinclair, A.R.E., J.M. Gosline, G. Holdsworth, C.J. Krebs, S. Boutin, J.N.M. Smith, R. Boonstra, and M. Dale. 1993. Can the solar cycle and climate synchronize the snowshoe hare cycle in Canada? Evidence from tree rings and ice cores. *Am. Nat.* 141:173–198. (10.2)

Singleton, G.R. and D.A. Hay. 1982. A genetic study of male social aggression in wild and laboratory mice. *Behav. Genet.* 12: 435–448. (12.1)

Smith, C.H. 1983. Spatial trends in Canadian snowshoe hare, *Lepus americanus*, population cycles. *Can. Field-Nat.* 97: 151–160. (10.2)

Smith, S.M. 1967. Seasonal changes in the survival of the Black-capped Chickadee. *Condor* 69: 344–359. (1.7)

Smyth, M. 1968. The effects of the removal of individuals from a population of bank voles *Clethrionomys glareolus*. *J. Anim. Ecol.* 37: 167–183. (9.3)

Soper, J.D. (1928). A faunal investigation of Southern Baffin Island. *Nat. Mus. Can., Biol. Ser.* No. 15, Bull No. 53: 1–143. (1.2)

Southern, H.N. 1954. *Control of rats and mice. Vol. 3. House mice.* Oxford: Clarendon. (4)

Southern, H.N. 1970. The natural control of a population of tawny owls (*Strix aluco*). *J. Zool. Lond.* 162: 197–285. (2.5)

Spears, N. and J.R. Clarke. 1987. Comparison of the gonadal response of wild and laboratory field voles to different photoperiods. *J. Reprod. Fertil.* 79: 231–238. (12.1)

Spears, N. and J.R. Clarke. 1988. Selection in field voles (*Microtus agrestis*) for gonadal growth under short photoperiods. *J. Anim. Ecol.* 56: 61–70. (6.3, 12.1)

Stenseth, N. C. 1981. On Chitty's theory for fluctuating populations: The importance of genetic polymorphism in the generation of regular density cycles. *J. Theor. Biol.* 90: 9–36. (1.4, 7.5)

Stenseth, N.C. and R.A. Ims, eds. 1993. *The biology of lemmings.* Symp. Linn. Soc. Ser. 15: 1–683 + xv. London: Academic Press, (12.2, 12.5)

Stenseth, N.C. and R.A. Ims. 1993. The history of lemming research: From the Nordic sagas to *The biology of lemmings*. In Stenseth and Ims, eds. *The biology of lemmings* Symp Linn. Soc. Ser. 15, pp. 4–34. (1.4, 7.2)

Stenseth, N.C. and W.Z. Lidicker, Jr. 1992. Presaturation and saturation dispersal 15 years later: Some theoretical considerations. In N.C. Stenseth and W.Z. Lidicker, Jr., eds *Animal dispersal: Small mammals as a model*, pp. 201–223. London: Chapman and Hall. (1.7, 9.3)

Stenseth, N.C. and W.Z. Lidicker, Jr., eds. 1992. *Animal dispersal: Small mammals as a model.* London: Chapman and Hall.

Stevens, H.M. 1937. A study of the vascularity of the pituitary body in the cat. *Anat. Rec.* 67: 377–94. (5.1)

Summerhayes, V.S. 1941. The effect of voles (*Microtus agrestis*) on vegetation. *J. Ecol.* 29:14–48. (2.2, 2.4)

Sutton, G.M. and W.J. Hamilton, Jr. 1932. The mammals of Southampton Island. *Mem. Carnegie Mus.* 12 (2): 3–107. (1.2, 5.8)

Szent-Györgyi, A. 1964. Teaching and the expanding knowledge. *Science* 146: 1278–1279. (Ack)

Taitt, M.J. and C.J. Krebs. 1985. Population dynamics and cycles. In R.H. Tamarin, ed. *Biology of New World* Microtus, Amr. Soc. Mammal. Spec. Publ. No. 8, pp. 567–620. (5.3)

Tamarin, R.H. 1977. Demography of the beach vole (*Microtus breweri*) and the meadow vole (*Microtus pennsylvanicus*) in southeastern Massachusetts. *Ecology* 88: 1310–1321. (12.5)

Tamarin, R.H. 1978. A defense of single-factor models of population regulation. In D. Snyder, ed. *Populations of small mammals under natural conditions, Pymatuning Lab. Ecol. Spec. Publ.*, pp. 159–162. (12.2)

Tamarin, R.H. and C.J. Krebs. 1969. *Microtus* population biology. II. Genetic changes at the transferrin locus in fluctuating populations of two vole species. *Evolution* 23: 183–211. (12.1, 12.4)

Tamarin, R.H. and C.J. Krebs 1973. Selection at the transferrin locus in cropped vole populations. *Heredity* 30: 53–62. (12.1)

Tamarin, R.H., L. Reich, and C.A. Moyer. 1984. Meadow vole cycles within fences. *Can. J. Zool.* 62: 1796–1804. (12.5)

Tamarin, R.H. and M. Sheridan. 1987. Behavior-genetic mechanisms of population regulation in microtine rodents. *Am. Zool.* 27: 921–927. (12.5)

Taubes, G. *Bad science: The short life and weird times of cold fusion.* New York: Random House. (P)

Thorpe, W.H. 1974. David Lambert Lack. *Biographical Mem. Fellows R. Soc.* 20: 271–293. (7.3)

Tinbergen, N. 1957. The functions of territory. *Bird Study* 4: 14–27. (8.9)

van Wijngaarden, A. 1960. The population dynamics of four confined populations of the continental vole *Microtus arvalis* (Pallas). *Versl. Landbouwkd. Onderz.* No. 66.22. (6.5)

Varley, G.C. 1949. Population changes in German forest pests. *J. Anim. Ecol.* 18: 117–122. (11.1)

Varley, G.C, and G.R. Gradwell. 1960. Key factors in population studies. *J. Anim. Ecol.* 29: 399–401. (11.1)

Varley, G.C., G.R. Gradwell, and M.P. Hassell. 1973. *Insect population ecology: An analytical approach.* Oxford: Blackwell. (11.1)

Verner, J. 1977. On the adaptive significance of territoriality. *Am. Nat.* 111: 769–775. (7.3)

Vogel, S. 1992. *Vital circuits: On pumps, pipes, and the workings of circulatory systems.* New York: Oxford University Press. (P, 1.1, 5.4, 12.1)

Voipio. P. 1950a. Evolution at the population level with special reference to game animals and practical game management. *Papers on Game Research* 5: 1–176. (1.7, 7.5, 8.5, 11.1, 12.1)

Voipio, P. 1950b. Jaksoittainen runsaudenvaihtelu ja paikallisten eläinkantojen sailyminen. English summary, pp. 162–64: On survival during cycles. *Suomen Riista* 5: 144–164. (1.7, 7.5, 8.5, 12.1)

Voipio, P. 1988. Comments on the implication of genetic ingredients in animal population dynamics. *Ann. Zool. Fennici* 25: 321–333. (1.7, 11.1, 12.1)

von Wright, G.H. 1957. *The logical problem of induction.* 2nd ed. Oxford: Blackwell. (3.1)

Wake, M.H. 1993. Two-career couples—attitudes and opportunities. *BioScience.* 43: 238–240. (5.1)

Warkowska-Dratnal, H. and N.C. Stenseth. 1985. Dispersal and the microtine cycle: Comparison of two hypotheses. *Oecologia* 65: 468–477. (1.4)

Watkins, J.W.N. 1964. Confession is good for ideas. In D. Edge, ed. *Experiment: A series of scientific case histories . . .* London: BBC. (1.1)

Watson A. and Moss, R. 1970. Dominance, spacing behaviour and aggression in relation to population limitation in vertebrates. In A. Watson, ed. *Animal populations in relation to their food resources,* pp. 167–218. Oxford: Blackwell. (9.3)

Watson, A., R. Moss, R. Parr, M.D. Mountford, and P. Rothery. 1994. Kin landownership, diffferential [sic] aggression between kin and non-kin, and population fluctuations in red grouse. *J. Anim. Ecol.* 63: 39–50. (11.1, 12.5)

Watson, J.D. and F.H.C. Crick. 1953. Molecular structure of deoxyribonucleic acid. *Nature.* 171: 964–967. (8.5, 11.2)

Wellington, W.G. 1957. Individual differences as a factor in population dynamics: The development of a problem. *Can. J. Zool.* 35: 193–323. (11.1)

Wells, A.Q. 1937. Tuberculosis in wild voles. *Lancet* 22 May, p. 1221. (2.8)

Wells, A.Q. 1946. The murine type of tubercle bacillus (the vole acid-fast bacillus). *Spec. Rep. Ser. Med. Res. Council, Lond.* 259: 1–42. (2.8)

Wells, A.Q. 1949. Vaccination with the murine type of tubercle bacillus (vole bacillus). *Lancet* 9 July 1949: 53–55. (2.8)

Whitehead, A.N. 1911–1926. In A.H. Johnson, ed. 1947. *The wit and wisdom of Whitehead.* Boston: The Beacon Press. (8.5)

Whitehead, A.N. 1942. *Adventures of ideas.* Harmondsworth, Middlesex: Penguin Books. (P, 1.4, 1.8)

Whitehead, A.N. 1950. *The aims of education and other essays.* 2nd ed. London: Ernest Benn Ltd. (1.1, 2.1, 3.1, 11.1, 11.2, 12.5)

Whitehead, A.N. 1953. Science and the modern world. Cambridge: Cambridge U P. (Ack, 1.1, 1.4, 1.5, 4, 5.8, 8.2. 8.5, 8.6, 8.7, 9.4, 10.2, 11.1, 11.2, 12.1)

Wiens, J.A. 1966. On group selection and Wynne-Edwards' hypothesis. *Am. Sci.* 54: 273–287. (8.9)

Wigglesworth, V.B. 1961. Insect polymorphism—a tentative synthesis. In J.S. Kennedy, ed., *Insect polymophism.* Symp. R. Ent. Soc. Lond. No. 1: 103–113. (1.6)

Wikander, R. 1985. Parsimony and testability: A reply to Dunbar. *Can. J. Zool.* 63:728–732. (P, 1.1)

Williams, G.C. 1966. *Adaptation and natural selection: A critique of some current evolutionary thought.* Princeton, NJ: Princeton University Press. (8.9)

Williams, G.R. 1954. Population fluctuations in some northern hemisphere game birds (Tetraonidae). *J. Anim. Ecol.* 23: 1–34. (8.1, 10.1).

Williams, J. 1985. Statistical analysis of fluctuations in red grouse bag data. *Oecologia* 65: 269–272. (10.1)

Williamson, M. 1972. *The analysis of biological populations.* London: Edward Arnold. (1.3)

Wolff, J.O. 1985. The effects of density, food, and interspecific interference on home range size in *Peromyscus leucopus* and *Peromyscus maniculatus*. *Can. J. Zool.* 63: 2657–2662. (12.5)

Wolff, J.O., M.H. Freeberg, and R.D. Dueser. 1983. Interspecific territoriality in two sympatric species of *Peromyscus* (Rodentia: Cricetidae). *Behav. Ecol. Sociobiol.* 12: 237–242. (12.5)

Wynne-Edwards. V. 1959. The control of population density through social behaviour: A hypothesis. *Ibis* 101: 436–441. (8.9).

Wynne-Edwards, V. 1962. *Animal dispersion in relation to social behaviour.* Edinburgh: Oliver and Boyd. (8.9)

Wynne-Edwards, V. 1991. Ecology denies Neo-Darwinism. *Ecologist* 21: 136–141. (8.9)

Ydenberg, R.C. 1987. Nomadic predators and geographical synchrony in microtine population cycles. *Oikos* 50: 270–272. (10.1)

Index

Authors are indexed by section number (in parentheses) in the references; but some of those quoted from Beveridge (1950) and Judson (1979) are also listed below.